Damage and Failure of Rock Triggered by Dynamic Disturbance

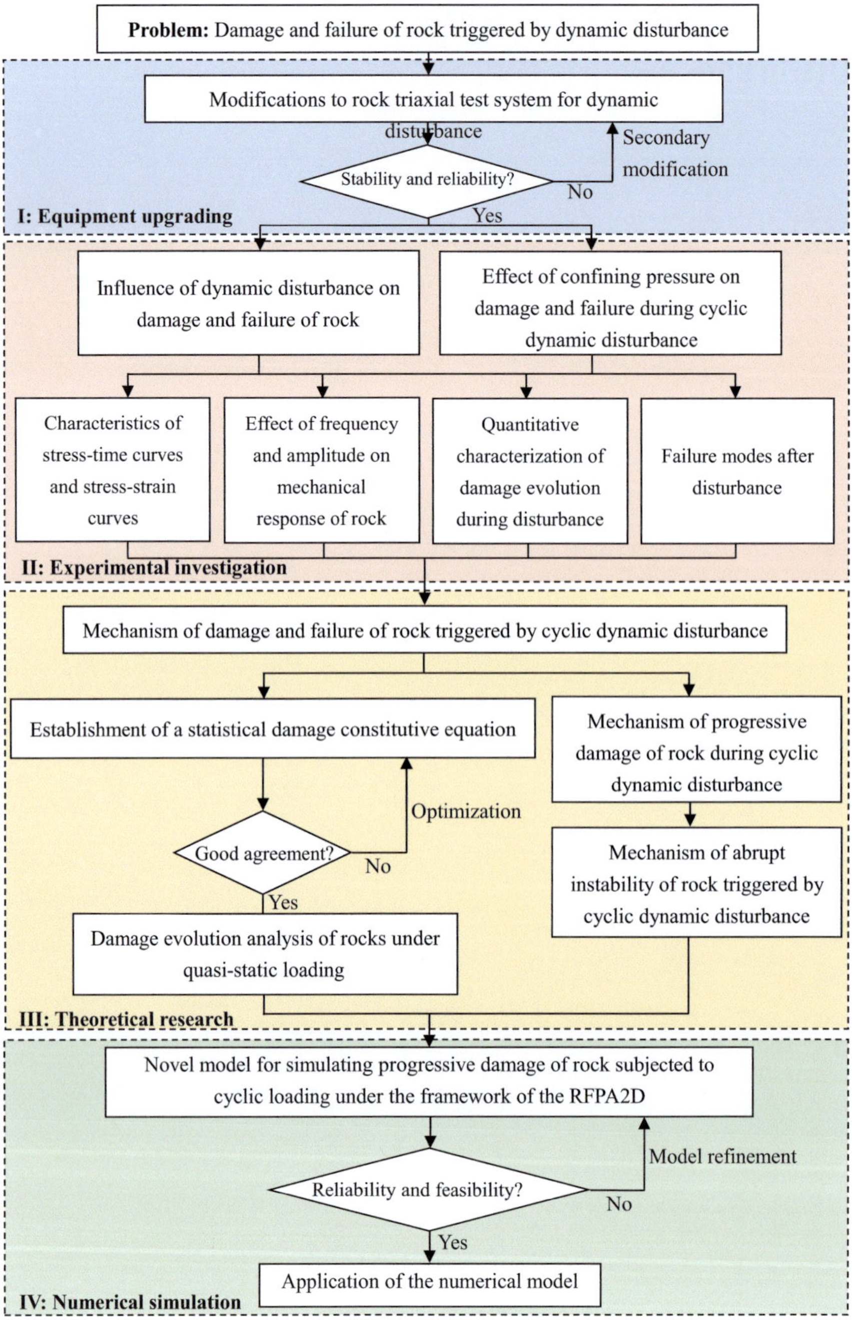

Problem: Damage and failure of rock triggered by dynamic disturbance
Modifications to rock triaxial test system for dynamic disturbance
Stability and reliability?
No
Secondary modification
Yes
I: Equipment upgrading
Influence of dynamic disturbance on damage and failure of rock
Effect of confining pressure on damage and failure during cyclic dynamic disturbance
Characteristics of stress-time curves and stress-strain curves
Effect of frequency and amplitude on mechanical response of rock
Quantitative characterization of damage evolution during disturbance
Failure modes after disturbance
II: Experimental investigation
Mechanism of damage and failure of rock triggered by cyclic dynamic disturbance
Establishment of a statistical damage constitutive equation
Good agreement?
No
Optimization
Yes
Damage evolution analysis of rocks under quasi-static loading
Mechanism of progressive damage of rock during cyclic dynamic disturbance
Mechanism of abrupt instability of rock triggered by cyclic dynamic disturbance
III: Theoretical research
Novel model for simulating progressive damage of rock subjected to cyclic loading under the framework of the RFPA2D
Reliability and feasibility?
No
Model refinement
Yes
Application of the numerical model
IV: Numerical simulation

Xu Chen

Damage and Failure of Rock Triggered by Dynamic Disturbance

Springer

Xu Chen
State Key Laboratory of Geohazard
Prevention and Geoenvironment Protection
Chengdu University of Technology
Chengdu, Sichuan, China

ISBN 978-981-96-9485-3 ISBN 978-981-96-9486-0 (eBook)
https://doi.org/10.1007/978-981-96-9486-0

This Springer imprint is published by the registered company Springer Nature Singapore Pte Ltd.
The registered company address is: 152 Beach Road, #21-01/04 Gateway East, Singapore 189721, Singapore

To my wife and little son, Rui Song and Shuke Chen, for their unwavering support and selfless love.

Preface

When excavation is carried out under high stress environment, the external disturbance will change the stress field of the rock, which may lead to potential engineering damage such as large deformation and large displacement of the free surfaces of the engineering surrounding rock and even cause instability damage, rockburst and other engineering disasters. From 1960s onwards, the subject of progressive damage and failure of rock materials has been extensively investigated to prevent and control geological disasters and engineering accidents. However, the existing researches usually did not take the effects of low-frequency dynamic disturbances, such as mechanical rock breaking vibration, remote blasting vibration, heavy machine vibration, on the mechanical properties of rock into consideration. In fact, the mechanism of rock damage and failure triggered by coupling effects of cyclic disturbance and high stress is very complicated, and common understanding has not yet been formed. Therefore, finding out the mechanisms of high-stress rock triggered by cyclic disturbance can be conducive to solving many stability problems in rock engineering.

The purpose in writing this book is to elucidate the mechanism of damage and failure of rock materials triggered by cyclic dynamic disturbance. Chapter 1 introduces the geological background of engineering failure and state of the art. Chapter 2 describes the preliminary development and secondary modification of dynamic disturbance rock servo triaxial test system in detail. Chapters 3–5 provide experimental results on the subject of the effects of cyclic dynamic disturbance and confining pressure on progressive damage and failure of rock. Chapter 6 analyzes the mechanism of damage and failure of rock triggered by cyclic dynamic disturbance from the perspective of theoretical research. Chapter 7 proposes a numerical method of rock damage triggered by cyclic dynamic disturbance. Embedded the numerical method into RFPA2D software, a new software named RFPA2D-Cycle is developed and the physical tests are numerically reproduced by employing it. Chapter 8 gives concluding remarks of this research and the suggestions for further research work.

The author hopes that the objective will have been achieved and that you will enjoy reading this book. If not, the fault is entirely the author's. Due to the limited scientific research level and lack of research ability of the author, the views and conclusions in this book and the prospect of the future research direction may be insufficient. The author sincerely invites experts and readers to criticize and correct them.

Chengdu, China Xu Chen

Acknowledgements This work was financially supported by the National Natural Science Foundation of China (42107208), the Natural Science Foundation of Sichuan Province (2024NSFSC0099), the Sichuan Science and Technology Program (2021YJ0394) and the State Key Laboratory of Geohazard Prevention and Geoenvironment Protection Independent Research Project (SKLGP2023Z014). The author is grateful for these supports.

Competing Interests The author has no competing interests to declare that are relevant to the content of this manuscript.

Contents

About the Author

Dr. Xu Chen was born in August 1988. In September 2015, he started his Ph.D. research, in Northeastern University, Shenyang, PR China, and got Mining Engineering of Doctor's degree in October 2019. From May 2020, he commenced his career in the College of Environment and Civil Engineering, Chengdu University of Technology, China, as a full-time lecture. From December 2022, he has been promoted as an assistant professor. In March 2023, he was elected as a full research staff of State Key Laboratory of Geohazard Prevention and Geoenvironment Protection (SKLGP), Chengdu University of Technology, China.

In the past five years, he was the young Principal Investigator (PI) of several key scientific projects including National Natural Science Foundation of China (NSFC), Sichuan Science and Technology Program, Natural Science Foundation of Sichuan Province, State Key Laboratory of Geohazard Prevention and Geoenvironment Protection Independent Research Project, etc. His current research interests focus mainly on damage and failure of rock triggered by cyclic dynamic disturbance, permeability evolution and progressive damage of rock and deep underground rock mass. Throughout his career, Dr. Xu Chen has published more than 30 papers in prestigious journals, in which more than 20 English papers (indexed by SCI) were published in *Rock Mechanics and Rock Engineering*, *Landslides*, *Construction and Building Materials*, *Environmental*

Earth Sciences, *Journal of Mountain Science*, *KSCE Journal of Civil Engineering*, *Tehnicki vjesnik—Technical Gazette*, *Journal of Zhejiang University-SCIENCE A* (Applied Physics and Engineering), *Journal of Central South University*, *Structural Engineering International*, etc.

Chapter 1
Introduction

1.1 Background

In the long history of human beings transforming nature to improve their living standards, rocks are one of the oldest and most commonly used geological materials on the earth. For example, primitive people used stone caves to shelter from wind and rain, and ground stones into hoes, shovels and other tools for farming. Ancient people also used stones to build castle fortifications, and used stone slates and beams to pave roads and build bridges. In today's highly developed science and technology, the transformation and utilization of rock materials are related to all aspects of the national economy, such as exploration of oil and gas (Sun et al. 2015), mining of minerals and coal (Zamani et al. 2023), construction of mountain tunnels (Home 2016), slope reinforcement of highways, railways and hydraulic engineering (Chen et al. 2016), prediction and early warning of geological disasters (Tang et al. 2015; Bhattacharya et al. 2012), development and utilization of geothermal resources (Pruess 2006; Ghassemi 2012), landfill and disposal of garbage, industrial waste, high-level radioactive nuclear waste (Hunsche and Hampel 1999; Hudson et al. 2001; Wang et al. 2018), etc. For some deep rock engineering, the surrounding rock is often in a coupling state of one or more fields among high stress fields, high temperature fields, and high permeability fields, and is even affected by other factors such as chemical corrosion (Tsang et al. 2012; Izadi and Elsworth 2015).

In actual engineering, mechanical rock drilling, remote blasting, continuous low-frequency vibration of heavy machinery, natural earthquakes, etc. often cause disturbance to the surrounding rock. These disturbance loads have the characteristics of long propagation distance and high carrying energy. Under the action of these disturbance loads, potential engineering hazards such as large deformation and large displacement (such as surface bulging, flaking, fracture and relaxation, etc.) may occur on the free surface of the surrounding rock under high ground stress conditions. Disasters such as mine earthquakes and rock bursts may even occur when some rock fragments fall off, collapse, and eject from the parent rock, causing strain

X. Chen, *Damage and Failure of Rock Triggered by Dynamic Disturbance*,
https://doi.org/10.1007/978-981-96-9486-0_1

energy to be instantaneously released from the rock. For example, the Dongguashan Copper Mine in Anhui Province, PR China, is buried at a depth of more than 1000 m. Rockburst accidents have occurred many times, threatened the safety of engineering personnel and equipment, and affected the mining progress. Another example is the diversion tunnel project of the Jinping II hydropower station in southwestern China. During the excavation process using a tunnel boring machine (TBM), medium and strong rock bursts occurred many times. The most intense one occurred at Stake SK9+283–9+322 in the drainage tunnel at 0:54 on November 28, 2009, with the depth of the blast pit of 8–9 m. The TBM machine was totally destroyed and buried under more than 400 m^3 of rock fragments to a distance of about 30 m behind the cutter head (Zhang et al. 2013). A seismic event with a Richter magnitude of 2.0 was detected. In this accident, the rock bolts fell off within 28 m of the tunnel depth, the support system was damaged, 7 workers were killed and 1 worker was injured.

Table 1.1 lists some rockbursts accidents occurred on mining areas and tunnel projects in PR China (Qian 2012). Judging from the conditions for the occurrence of these engineering disasters, most of the projects are located in relatively large burial depths and relatively high ground stresses, and all of them occur due to the interaction between the surrounding rock and the high stress during excavation. Therefore, how to carry out engineering design and construction, prevent and control engineering disasters, and ensure the stability and safety of engineering surrounding rock under the combined action of external disturbance and high ground stress is an urgent problem to be solved.

At present, scholars generally believe that two conditions, high stress environment and external disturbance, need to be met simultaneously for the occurrence of rock burst (Walton et al. 2015; Yang et al. 2017; Fu et al. 2020). Specifically, when underground tunnels, ore pillars, etc. are in a high-geostress environment, external disturbances will change the stress field conditions of rocks, thereby inducing large deformation and cumulative damage of rocks, and even causing engineering disasters such as instability damage and rock bursts. However, most existing literature uses static methods to study the rock excavation process. The effect of cyclic disturbances such as mechanical rock breaking vibration, remote blasting vibration, heavy machinery vibration, mine collapse vibration, natural earthquakes, etc. on rock mechanical properties were usually not taken into consideration. Until now, the mechanism of rock damage and failure triggered by low-frequency cyclic disturbances under high stress is very complex, and there is still no unified understanding yet. Therefore, studying the deformation, damage and failure of rocks under high stress conditions triggered by cyclic disturbances, and exploring the relevant mechanisms in this process will help solve many problems existing in the stability of rock engineering, and it will also have positive significance for improving rock breaking technology and support technology.

Table 1.1 Some rockbursts accidents occurred on mining areas and tunnel projects in PR China (Qian 2012)

Project	Completion time	Maximum buried depth/m	Rockburst intensity and proportion/%			Number of rock bursts	Impact length/m
			Slight	Medium	Strong and extremely strong		
Guancunba Tunnel of Chengdu-Kunming Railway	1996	1650	Main	Small amount	None	–	–
Diversion Tunnel on the left bank of Ertan Hydropower Station	1993	200	Main	Small amount	None	–	315
Diversion Tunnel of Minjiang Taipingyi Hydropower Station	1993	600	Main	Small amount	None	> 400	–
Diversion Tunnel of Tianshengqiao II Hydropower Station	1996	800	70.0	29.5	0.5	30	–
Qinling Railway Tunnel	1998	1615	59.3	34.3	6.4	–	1894
Erlangshan Tunnel on the Sichuan-Tibet Highway	2001	760	Main	Small amount	None	> 200	1252
Chongqing Tongyu Tunnel	2002	1050	91.0	7.8	1.2	–	655
Chongqing Lujialing Tunnel	2004	600	55.8	39.7	4.5	93	–
Pugou Hydropower Station Inlet Traffic Tunnel	2005	420	–	–	–	183	–
Qinling Zhongnanshan Highway Tunnel	2007	1600	61.7	25.6	12.7	–	2664
Diversion Tunnel of Jinping II Hydropower Station	2011	2525	44.9	46.3	8.8	> 750	–

(continued)

Table 1.1 (continued)

Project	Completion time	Maximum buried depth/m	Rockburst intensity and proportion/%			Number of rock bursts	Impact length/m
			Slight	Medium	Strong and extremely strong		
Diversion Tunnel of Jiangbian Hydropower Station	2012	1678	46.4	50.4	3.2	> 300	–
Dongguashan Copper Mine	–	1007	Main	Small amount	None	> 100	85

1.2 State of Art

In rock engineering, engineering surrounding rock is often affected by external disturbances such as mechanical rock breaking, rock bursts, earthquakes, slab cracking, and high-stage ore falling. For some rock engineering in high-stress environments, some dynamic disasters are often induced during the excavation process, and the disturbance of these dynamic disasters can react on the confining pressure of the near-end or far-end engineering, further amplifying the disaster. Therefore, from the perspective of stress conditions, engineering rocks (or rock mass) are often in a combined stress state of "critical static stress þ slight disturbance" and "elastic static stress + impact disturbance" (Li et al. 2017). This stress state is somewhat different from the stress state of rocks in shallow engineering. Accordingly, it is difficult to use the classical rock mechanics theory established by shallow rock mechanics research to explain the essential causes and internal mechanisms of some disasters in deep rock under the combined stress state of external disturbance and high ground stress. Figure 1.1 depicts the stress state of the rock mass on the free surface during the excavation process of the deep tunnel (Su et al. 2017). As can be seen from the figure, during excavation by means of full-face tunnel boring machine (TBM) or borehole-blasting method, there are usually drilling, blasting, mechanical rock drilling and other processes on the tunnel face. These disturbances will use the tunnel surrounding rock as the medium and be transmitted to the free surface in the form of stress waves. The surrounding rock of deep tunnels is often in a three-dimensional stress state, that is, the circumferential, axial and radial directions of the tunnel are all subjected to in-situ stress. One of the rock units on the free surface is a free surface, which allows a larger range of deformation. When subjected to static high ground stress, the elements on the free surface can often maintain a stable state. However, once it is disturbed by external disturbances from the far field, the original stable state will be destroyed, resulting in sudden catastrophic phenomena [such as rock burst (Zhu et al. 2010; Zhang et al. 2013; Feng et al. 2019), layer cracks (Beom et al. 2013; Zhou et al. 2018)]. These catastrophic phenomena may further induce catastrophic

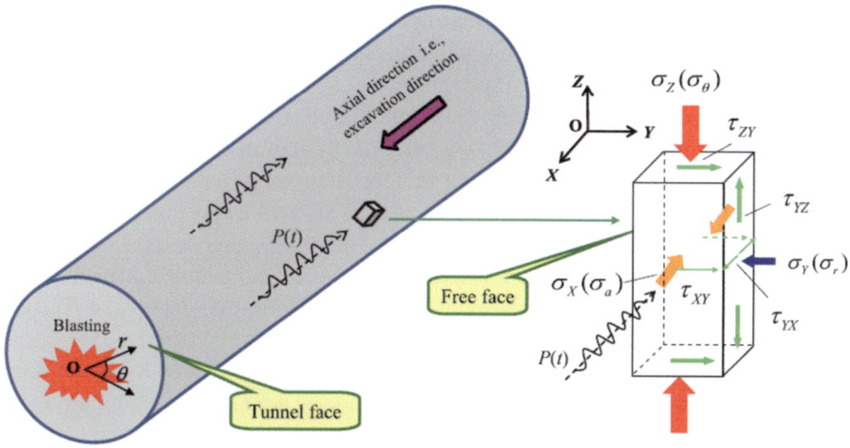

Fig. 1.1 Stress state of rock masses on free surface during excavation (Su et al. 2017)

disasters in other locations in the tunnel in the form of disturbances. In this context, many scholars have begun to conduct research on the mechanical response, deformation and failure characteristics, and damage evolution rules of rocks under cyclic loads or external disturbances, and have tried to explore the essential reasons for the instability and failure of deep rock engineering under this stress mode. In view of this, the following is a brief review of previous relevant research from two aspects: the mechanical properties and damage characteristics of rocks under the action of cyclic load disturbance, and the mechanical properties of rocks under combined dynamic and static loading.

1.2.1 Mechanical Properties and Damage Mechanism of Rock Under Cyclic Disturbance Loading

In deeply buried rock engineering, mechanical drilling, remote blasting, continuous low-frequency vibration of heavy machinery, and natural earthquakes often cause cyclic disturbance of the surrounding rocks. These external disturbances are usually more complex in terms of frequency, amplitude, waveform and other information, and are a very complex form of loading in rock engineering. In order to facilitate research, many scholars directly simplify these complex load forms into a cyclic form of alternating loading and unloading during the process of indoor experimental research. They believe that the damage and destruction of rocks caused by external disturbances are the result of cyclic loading. A large number of studies have found that rock samples will show irreversible deformation (strain) during the cyclic loading and unloading process, so that the damage caused by cyclic loading and unloading of rock is considered to be a repetitive accumulation of irreversible deformation in a single cycle, and it is considered that the fatigue damage of rocks under cyclic loading is the essence of the failure of rocks subjected to cyclic disturbance. For example, Ge et al. (2003) and Jiang et al. (2004) conducted uniaxial compression, cyclic loading and unloading tests on rocks, and found that the damage of rock under cyclic loading and unloading is associated with the complete stress–strain curve under triaxial compression. They found that the deformation amount when failure occurs is equivalent to the strain of the post-peak stress–strain curve corresponding to the upper limit stress of cyclic loading. Taylor et al. (1986) analyzed the cumulative damage of microcracks in rocks under cyclic loading and unloading, and established a theoretical damage model for materials exhibiting strain-rate-dependent fracture behavior.

External disturbance loads acting on rocks will definitely cause rock deformation, so many scholars take the study of the strength and deformation of rocks under cyclic loading and unloading as a starting point. In the 1970s, Attewell and Farmer (1973) conducted fatigue tests on dolomite and proposed a hypothesis that crack expansion related to strain energy leads to rock deformation and damage. During the same period, Lin and Wu (1976) also used a rock rigidity testing machine to conduct cyclic loading and unloading tests on rocks. They found that after repeated cyclic loading and unloading of rocks, the stress–strain curve would still develop along the monotonic loading process and would be less affected by cyclic loading. Subsequently, Lin and Wu (1987) studied the strength and deformation properties of two kinds of rocks (Hubei marble and Sichuan red sandstone) under cyclic loading by using MTS 815 electro-hydraulic servo tester. The results showed that the strength of both types of rocks under cyclic loading is lower than that under the proposed static loading, and the cyclic fatigue strength limit is directly related to the amplitude and frequency, etc., whereas the deformation nature of rocks under cyclic loading is very similar to that of creep deformation that undergoes the action of the proposed static loading. Jin et al. (1992) conducted triaxial compression cyclic loading and unloading tests on marble under different confining pressures, and discussed the causes of rock deformation, instability and failure during cyclic loading and unloading from the perspective of the testing machine-rock sample interaction system. They also found that the failure mode of rock under cyclic loading and unloading is different from conventional loading. Singh (1989) studied the fatigue and strain hardening behavior of Australian Flagstaff limestone samples under cyclic loading and found that the fatigue life of limestone increases with the decrease of stress amplitude, and as the stress amplitude decreases, the fatigue life of limestone increases. The number of cycles increases logarithmically. Zhang et al. (2004) observed the deformation and acoustic emission characteristics of rocks under cyclic loading, analyzed the axial and transverse deformation patterns of rocks from macroscopic and microscopic perspectives, and predicted the precursor phenomena of rocks before macroscopic rupture occurs.

Some other scholars have begun to use the cyclic loading and unloading load application method to conduct detailed research on the stress–strain relationship, deformation characteristics, residual strength, failure characteristics, energy characteristics, damping ratio, etc. of rocks. Taheri et al. (2016) conducted triaxial monotonic compression tests and cyclic loading and unloading tests with different stress levels and unloading amplitudes on Hawkesbury sandstone samples. The results showed that the brittleness of rock is related to the number of cyclic loading and unloading cycles experienced, and the mechanical properties of the rock change with the stress level of the applied cyclic loading. If the applied cyclic loading and unloading stress is lower than the critical deviatoric stress which is in the range of 93.7–94.1% of the maximum deviatoric stress, the ultimate compressive strength of the rock sample will increase, and the increase in strength will be proportional to the increase in cyclic loading and deviatoric stress. Tao and Mo (1990) conducted two different types of cyclic loading and unloading tests on E-Cheng sandstone and Da-Ye marble, and some factors, such as stress levels, cyclic load waveforms, and amplitude, etc., that affect the deformation patterns of rock samples were discussed. Wang et al. (2013) performed cyclic loading and unloading tests under triaxial compression conditions to study the effects of volume change and residual strain on the deformation modulus of granite. They found that rocks have three different fatigue behavior domains in the stress–strain space, namely volume compression, strain-strengthened volume expansion, and strain-softening volume expansion. By a suit of conventional triaxial tests and cyclic loading–unloading tests on red sandstone under different confining pressure conditions, Zhang et al. (2006) found that the axial strain of red sandstone specimens when fatigued under cyclic loading is equivalent to the axial strain corresponding to the upper stress horizontal line after the peak of the full stress–strain curve. Yang et al. (2007) designed a set of uniaxial cyclic loading and unloading tests to study the strength, deformation and damage process of coal and rock collected from Baodian Coal 3, and found that the failure process of coal rock under cyclic loading can reflect the damage evolution process from compaction, strain hardening to softening. Wang et al. (2010) carried out cyclic loading and unloading tests with different confining pressures and different unloading degree on Sichuan red sandstone samples that were in a state of residual strength. It was found that under the same unloading degree, lower confining pressures will cause a greater reduction in the residual strength of rock samples under cyclic loading and unloading. Based on the test results of uniaxial cyclic loading and unloading of water-saturated diorite in the deep Zhongguan Iron Mine, Xia et al. (2014) used the theory of linear elastic damage mechanics to study the deformation, damage and energy consumption characteristics of saturated rock under the action of uniaxial cyclic loading and unloading. Liu et al. (2010), Liu and He (2012) conducted uniaxial compression cyclic loading and unloading tests on fine sandstone and silty mudstone to test the damping parameters, and discussed the relationship between rock density and cyclic loading characteristics, such as plastic deformation, hysteresis loop area, damping ratio, damping coefficient, etc. Shahnazari et al. (2010) tried to establish a numerical model—multi expression programming (MEP)—for modeling stress–strain behavior of sand under cyclic loading, which are particularly valuable for providing an analysis tool accessible to practicing engineers.

At the same time, some scholars studied the effect of load parameters during the cyclic loading and unloading process, such as the waveform (Bagde and Petroš 2005a), frequency (Bagde and Petroš 2005b; Liu et al. 2012a; Deng et al. 2017), amplitude (Liu et al. 2012b; Zhang et al. 2014), stress level (Zhao et al. 2017; Chen et al. 2018) and other parameters (Zhao et al. 2014; Huang et al. 2017) related to the disturbance load, on mechanical properties of rocks. The predefined frequency range is from 0.0019 to 20 Hz, and the waveforms include triangle waves, sine waves, square waves, etc. The lithology includes marble, granite, limestone, sandstone, coal rock, gypsum, shale, rock salt, gabbro, etc., and the loading path includes constant amplitude and variable amplitude (as shown in Fig. 1.2). The previous studies also cover jointed and intact rock samples. Research shows that the frequency, amplitude, and waveform of loading and unloading directly affect the shape of the stress–strain hysteresis loop, and rocks with different lithologies produce different deformation responses to cyclic loading conditions.

The continuous advancement of testing technology and the continuous introduction of new equipment and new methods have also provided favorable conditions for the study of rock damage evolution and failure mechanisms under cyclic loading conditions. Many scholars have introduced acoustic emission (AE) instruments (Meng et al. 2016, 2018; Zhou et al. 2019; Jiang et al. 2019) or digital image correlation (DIC) technique (Zhang et al. 2012; Song et al. 2013) during the test process to systematically observe the damage characteristics and deformation field of rock during cyclic loading and unloading. Meng et al. (2016, 2018) carried out a uniaxial cyclic loading and unloading compression experiment to explore the AE characteristics of rock under different loading and unloading rates. It was observed that the AE characteristics are closely related to the stress–strain properties of rock materials and that they are affected by the developmental state and degree of internal microcracks. Zhou et al. (2019) performed cyclically triaxial tests on Beishan granite and found that the propagation of fracture can be divided into different stages according to the AE hits and events. Jiang et al. (2019) also used AE data for assessing the evolution of AE

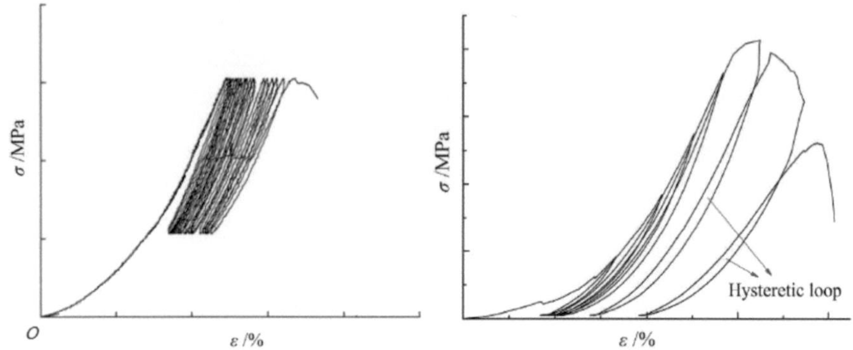

(a) constant amplitude (b) variable amplitude

Fig. 1.2 Two different types of cyclic loading paths

energy distribution and AE waiting time of sandstone under uniaxial cyclic loading. Zhang et al. (2012) obtained deformation fields on the surface of sandstone samples during cyclic indentation tests via DIC technique. Song et al. (2013) did DIC analysis on optical images of sandstone specimen surface in loading–unloading periods. The application of these new experimental technologies can intuitively observe the deformation, damage, and failure process during the cyclic loading and unloading process. Reports on experimental phenomena have deepened people's understanding of the impact of rock mechanical properties under cyclic disturbances, and provided experimental basis for qualitative research to quantitatively describe the rock damage process caused by cyclic loading–unloading disturbances.

Although it is difficult to quantitatively describe the rock deformation, damage, and failure processes caused by cyclic loading and unloading disturbances, many scholars have still conducted research in this area and achieved some enlightening results. For example, Xiao et al. (2010), Zhu et al. (2009), and Ni et al. (2011) used different physical quantities, such as residual strain angles, meso-mechanical angles, etc., to define damage variables on the basis of their cyclic loading–unloading tests on rocks. Their research results, which were further confirmed and promoted by Liu et al. (2012a, b), make the damage evolution during cyclic loading and unloading quantify. Li et al. (2001) conducted an experimental study on jointed rocks with loading frequencies of 0.2, 2, and 21 Hz, and load levels of 0.2–1 times the uniaxial compressive strength until failure. Based on the test results, they established a rock cyclic loading and unloading damage evolution model related to stress intensity factors under single joint and multi-joint conditions. Zhou et al. (2006) established a planar elastoplastic cellular automaton model for strain-softened rock materials based on the concept of mesomechanics. Through numerical simulation, they found that the uniaxial compressive stress–strain curve of rock materials simulated by this model was as follows: The cyclic loading and unloading stress–strain curves are consistent with the experimental results.

To sum up, the cyclic loading and unloading process of rock mechanics is a disturbance process to the rock in essence. The cracks number, crack density, crack connectivity of the rock change in this disturbance process. As a result, the damage of the rock continues to accumulate and the degree of damage continues to increase, eventually reaching the critical condition of failure, and losing bearing capacity.

Mechanical properties of rocks are greatly affected by cyclic external loads during cyclic loading and unloading. There are many influencing factors include waveform, frequency, amplitude, stress level, etc., and the cross-influence of each factor is complex. Therefore, it is difficult to quantitatively establish a constitutive relationship of rock suitable for cyclic loading and unloading using existing theories.

1.2.2 Mechanical Properties and Damage Mechanism of Rock Under Coupled Static-Dynamic Loads

The above briefly describes the relevant research results on the mechanical properties and damage mechanism of rocks under cyclic loading. The cyclic loading–unloading test is a method to further explore the mechanical properties and fatigue performance of rock using triaxial test machine. The repeatedly loading cycles can also be regarded as a form of disturbance caused by external loads acting on the rock. However, during this external disturbance process, the loading and unloading rates are often very small, and the rock sample is still in a quasi-static equilibrium state, which cannot fully reflect the stress state of rock damage and failure under high stress induced by external load disturbance. To this end, Professor Li Xibing's team from Central South University, PR China developed a rock dynamic and static combined loading test system (Li et al. 2008a, 2009, 2017; Zhou et al. 2014; Tao et al. 2017; Li et al. 2020), and realized the combined loading of "pre-static load + dynamic disturbance" in the true sense, which had very significant value in improving the basic data, theory and experimental methods of rock mechanics. Therefore, this section mainly will provide a brief review of relevant reports on the mechanics and damage characteristics of rocks under combined dynamic and static stress.

The strain rates of blasting and mechanical rock drilling range from approximately 100 s^{-1} (1 Hz) to 102 s^{-1} (100 Hz) or even higher. Many studies have shown that mechanical properties such as rock strength, deformation, damage evolution, and failure morphology are related to its loading rate. When it comes to higher strain rate loading, rock mechanics research often starts from the perspective of dynamics and adopts dynamics-related theories and methods. Specifically, when it comes to experimental research, separate Hopkinson pressure bars (SHPB) are often used to conduct impact tests on rock samples of a certain length, and the propagation of shock waves in the rock samples and the damage caused by the shock waves to the rock samples are checked. However, the traditional separated Hopkinson pressure bar used in rock dynamics research cannot realize the simultaneous application of static preloading and dynamic impact, and cannot be fully used to simulate the actual excavation conditions of the surrounding rock in the project. Therefore, Li et al. (2008a) modified the traditional separated Hopkinson pressure bar to apply axial prestress, axial impact load and confining pressure on the rock sample simultaneously, which is used to simulate the combined loading and stress state of "pre-static load + dynamic disturbance" in deep rocks. The maximum axial stress and maximum confining pressure that can be pre-applied to the modified separated Hopkinson compression bar (as shown in Fig. 1.3) are both 200 MPa. When conducting rock impact tests, the impact load strain rate range that can be applied is 10^2–10^3 s^{-1} (100–1000 Hz), which is a combination of dynamic and static loads with medium strain rates.

Using the modified separate Hopkinson pressure bar, the research team in Central South University conducted a large number of uniaxial and triaxial impact tests under different combined loading test conditions of pre-static load and impact disturbance. For example, Li et al. (2008b) conducted research on the strength characteristics,

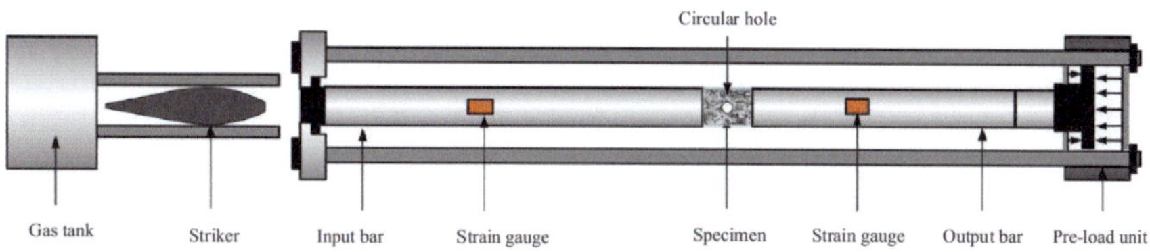

Fig. 1.3 A modified split Hopkinson pressure bar with the function of applying coupled dynamic and static loads developed at Central South University, PR China (Li et al. 2008a; Tao et al. 2017)

crushing rules and energy efficiency of rocks under different dynamic and static combined loading conditions. The results show that the main failure form of rock under combined dynamic and static loading is tensile failure. The increase of static load and dynamic load will cause the blockiness distribution of the broken rock to develop towards the fine-grained section, and the combined dynamic and static loading can effectively increase the energy utilization rate of the rock and improve the rock breaking effect. Zhou (2007) conducted experimental research on siltstone in two loading modes: "same dynamic impact load, different axial static loads" and "same axial static load, different impact loads". It is found that when the impact dynamic load is constant, the magnitude of the axial static load will affect the strength of the rock. When the axial static load is fixed, the impact dynamic load will cause the rock to show a strong strain rate effect. Gong et al. (2010, 2011) also used the modified equipment to conduct one-dimensional and three-dimensional dynamic and static combined loading dynamic characteristic tests on sandstone. The research results showed that the impact strength of the rock, the shape after the peak of the stress–strain curve, and the damage modes, etc. are all affected by the axial pressure ratio. At the same time, it is proved that the dynamic mechanism of rockburst occurs: a large amount of elastic strain energy stored inside is enough to cause the rock to be damaged and unstable. In the static or quasi-static state, the rock can remain stable, but after external dynamic disturbance loads are applied, cracks inside the rock are more likely to be activated, which induces a sudden release of elastic energy inside the rock, leading to rock instability and damage. Zhou et al. (2014) conducted a Brazilian splitting indirect tensile test on granite samples under combined dynamic and static loading. The test results showed that the tensile strength of granite decreased with the increase of pre-static load. Afterwards, Li et al. (2015a, b) performed Hopkinson compression bar tests on granite samples containing circular and square holes under different axial pre-static loads and the same impact dynamic load. Using nuclear magnetic resonance (NMR) tests before and after dynamic and static loading, they attempted to understand the microscopic structural damage rules of rocks containing holes under combined dynamic and static loading conditions. The test results showed that the greater the pre-imposed axial static load, the more sensitive the damage and degradation characteristics of the rock under impact are. The nuclear magnetic resonance image can more intuitively reflect the evolution law of the damage and degradation of the rock. Liu et al. (2014) used the improved Hopkinson pressure bar to conduct a one-dimensional dynamic failure test study of combined dynamic and static loading on combined coal and rock, and found that combined coal and rock with more fissures has stronger resistance to high dynamic loads and less severe damage. In the review article, Li et al. (2017) summarized his team's research results on the mechanical properties and damage mechanisms of rocks under combined dynamic and static loading. The modified Hopkinson compression bar made outstanding contributions to exploring the mechanical properties of rocks under the "pre-static loading + dynamic disturbance" mode. It provided insights into the mechanism of rockbursts and the micromechanical characterization of rock damage.

The modified split Hopkinson pressure bar developed by the research team at Central South University plays a very important role in studying the mechanical properties and damage mechanism of rocks under the combined loading of "pre-static load + dynamic disturbance". However, this device also has certain shortcomings. In this stress state of "pre-static load + dynamic disturbance", the external disturbance load is an impact load, which usually requires the use of dynamic methods for analysis. In actual rock engineering, during the deformation, damage and failure of rocks under high stress conditions induced by external disturbances, the external disturbance load may also be a "slight disturbance" with a relatively low frequency and a relatively small amplitude that causes a small strain rate in the rock, which is not enough to consider its dynamic properties. Therefore, Zuo et al. (2005, 2006) conducted research on the mechanical properties of rocks under the combined loading of "pre-static load + slight disturbance", which has a different loading mode from "pre-static load + dynamic disturbance". External slight disturbance in the loading mode of "pre-static load + slight disturbance" adopts low-cycle fatigue loading method, which belongs to the scope of statics research. In his doctoral thesis, Zuo (2005) introduced in detail the mechanical response and damage characteristics of red sandstone when subjected to "pre-static load + slight disturbance" with different pre-static loads and different slight disturbances under one- and two-dimension. Based on the experimental results, he explained the mechanism of rock damage and failure under one-dimensional static loading and combined dynamic and static loading from the aspects of elastic–plastic damage constitutive model, damage strain energy release rate, catastrophe theory stability analysis, etc.

The combined load of "pre-static load + dynamic disturbance" exerted by the improved Hopkinson compression bar is in the medium–high strain rate, and the combined load of "pre-static load + slight disturbance" is in the high strain rate. These two loading modes do not consider the true three-dimensional stress state of the rock (Du et al. 2014). In view of this, Du et al. (2015, 2016) and Li et al. (2015a, b) reported a TRW-3000 true triaxial testing equipment (as shown in Fig. 1.4) which can simulate the rock failure process under the combination of slight disturbance and high stress by applying a triaxial static load on the cubic rock sample with a slight disturbance. Through the combined loading test of "pre-static load + slight dynamic disturbance" under different paths, Du et al. (2016) found that the occurrence of rock burst is controlled by the amplitude of the disturbance load and the increasing static load. Only when the load exceeds a certain threshold, the elastic strain energy stored

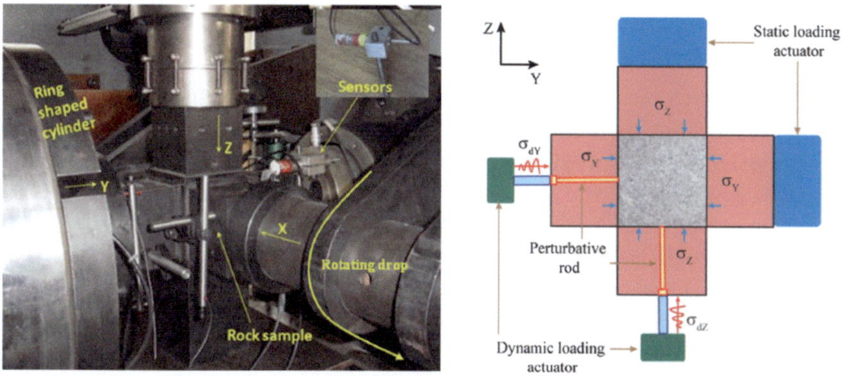

(a) Loading actuators along three directions (b) dynamically disturbed loading method

Fig. 1.4 Schematic diagram of rock true triaxial loading system using to apply "pre-static loading + slight dynamic disturbance" (Du et al. 2016)

in the rock sample will be suddenly released and rock fragments will fly out, that is, a rock burst will occur.

Moreover, Su et al. (2017) used a self-developed true triaxial rockburst test system to conduct experimental research on the rockburst process under the action of low-frequency periodic dynamic disturbance loads. The results show that when the amplitude of the disturbance load exceeds the threshold value, the possibility of rock burst in the rock sample increases sharply, and the frequency value of the disturbance load will directly affect the ejection kinetic energy during the rock burst process. Hu et al. (2018, 2023) used a new high-pressure servo dynamic triaxial rockburst testing machine to conduct true triaxial coupled static and dynamic loading tests, and studied the effects of static load tangential, axial and radial stress, dynamic disturbance amplitude and frequency on triggered rock bursts, and revealed the rock damage evolution rules and energy mechanisms of triggered rock bursts under different forms of dynamic disturbances. In addition to experimental studies, Zhu et al. (2012) also developed Real Fracture Process Analysis for Dynamics (RFPA-Dynamics) code to numerically simulate the failure process of rock under combined static and dynamic SHPB tests. The results show that the rock strength growth factor under combined dynamic and static loading is related to the homogeneity, static load level and strain rate. From the perspective of numerical simulation, this work reproduced the propagation law of stress waves in rock samples under combined dynamic and static loading conditions, provided the essential reasons for the damage and failure of rock samples caused by combined dynamic and static loading, and at the same time verified the relevant conclusions of experimental research. Therefore, compared with the traditional monotonic loading mode or cyclic loading–unloading mode, the combined loading mode of "pre-static load + dynamic disturbance" or "pre-static load + slight disturbance" can more truly reflect the actual stress state during deep rock excavation.

1.2.3 Limitation of Previous Studies

From the research results, the previous research has improved the basic theory, test method and technology of rock mechanics, and has a very important reference value for the design, construction and long-term safety of rock engineering (especially deep rock engineering). But there are still some shortcomings in these studies:

Firstly, in deep rock engineering, external cyclic dynamic disturbances actually work together with the high-stress environment of surrounding rock. When rocks are subjected to high ground stress, they are affected by external disturbances, resulting in deformation, damage or even failure and instability. Most previous studies have considered the influence of monotonic stress or cyclic load in isolation. However, the relevant mechanical response and mechanical properties of rock damage and failure under high stress conditions triggered by external cyclic dynamic disturbance are not taken into account.

Secondly, the synergistic effect of external disturbance and the high stress environment of rocks is considered in the combined dynamic and static loading test. However, there are still few studies on the damage and failure of rocks under high ground stress triggered by cyclic dynamic disturbance, and the mechanical response of rocks under high stress triggered by cyclic dynamic disturbance is not yet known. There is still no theoretical basis for the analysis of the evolution law and sudden instability in the process of rock damage and failure.

Thirdly, numerical simulation studies on rock damage and instability failure triggered by cyclic dynamic disturbance are still rarely reported, mainly because most of the current numerical simulation methods are based on the concept of elasticity. It is difficult for them to reflect the process of rock material gradually producing irrecoverable residual deformation and deteriorating strength under continuous cyclic dynamic disturbance loading.

1.3 Governing Flowchart for the Book

In view of these shortcomings, this study adopts the method of combining laboratory experiment, theoretical analysis and numerical simulation to investigation of damage and failure of rock triggered by dynamic disturbance. The governing flowchart for this book is constructed both as the Frontispiece and in Fig. 1.5. Firstly, a rock dynamic disturbance servo triaxial test system is independently developed and modified for cyclic dynamic disturbance loading. Secondly, the effects of cyclic dynamic disturbance on progressive damage and failure of rock and the confining pressure effects are tested, and the mechanical response, failure mode analysis and progressive damage and degradation process of rock triggered by cyclic dynamic disturbance are analyzed and discussed. Moreover, from the perspective of mathematics and mechanics, a constitutive model of rock damage evolution is constructed, and the phenomenon of rock deformation, damage and destruction under the action of high stress triggered by cyclic dynamic disturbance is explained. Finally, a numerical realization method of rock damage triggered by cyclic disturbance is proposed, and RFPA2D-Cycle is developed on the basis of RFPA2D software, and numerical simulation research is carried out on some laboratory tests.

1.4 Structure and Content of the Book

The book has been structured with seven further chapters, the titles and a brief summary of which are listed below.

In Chap. 2, **Modifications to Rock Triaxial Test System for Dynamic Disturbance**: The preliminary development and secondary modification of dynamic disturbance rock servo triaxial test system is described in detail. The stability and reliability of the developed equipment are also verified through a set of tests.

In Chap. 3, **Description of Rock Samples and Their Basic Physical and Mechanical Properties**: The basic physical properties of the rock sample are tested in terms of apparent density, porosity, S- and P-wave velocity of each rock sample before the triaxial cyclic dynamic disturbance tests.

In Chap. 4, **Experimental Study on the Influence of Dynamic Disturbance on Progressive Damage and Failure of Rock**: A series of disturbance loading tests with different disturbance frequencies and different disturbance amplitudes is carried out on sandstone samples at the axial stress level of 85% of uniaxial compressive strength. And the mechanical response, failure mode analysis and progressive damage and degradation process of rock triggered by cyclic dynamic disturbance are analyzed and discussed.

In Chap. 5, **Effect of Confining Pressure on Damage Evolution and Failure Behaviors of Intact Sandstone Samples during Cyclic Dynamic Disturbance**: A series of tests with a continuous axial cyclic loading in triaxial stress states are carried out under confining pressures ranging from 0 to 40 MPa to gain a better understanding of the effect of the confining pressure on the damage evolution and failure behaviors of intact sandstone samples during cyclic disturbance.

In Chap. 6, **Mechanism of Damage and Failure of Rock Triggered by Cyclic Dynamic Disturbance**: A statistical damage constitutive equation of rock in quasi-static loading phase is established, the nucleation phenomenon of microcracks in the cyclic dynamic disturbance phase is analyzed, and the abrupt instability of the rock sample at the end of cyclic dynamic disturbance phase is explained using cusp catastrophe theory.

In Chap. 7, **Numerical Model of Progressive Damage and Failure of Rock Triggered by Cyclic Dynamic Disturbance**: A novel assumption that the strength of each mesoscopic finite element degrades with cyclic numbers as a linear function and implant this assumption under the framework of RFPA2D is proposed. The reliability of the proposed model is examined by numerically performing cyclic loading tests for sandstone samples and is employed to investigate the damage progressivity of underground tunnels induced by the remote vertical cyclic loading.

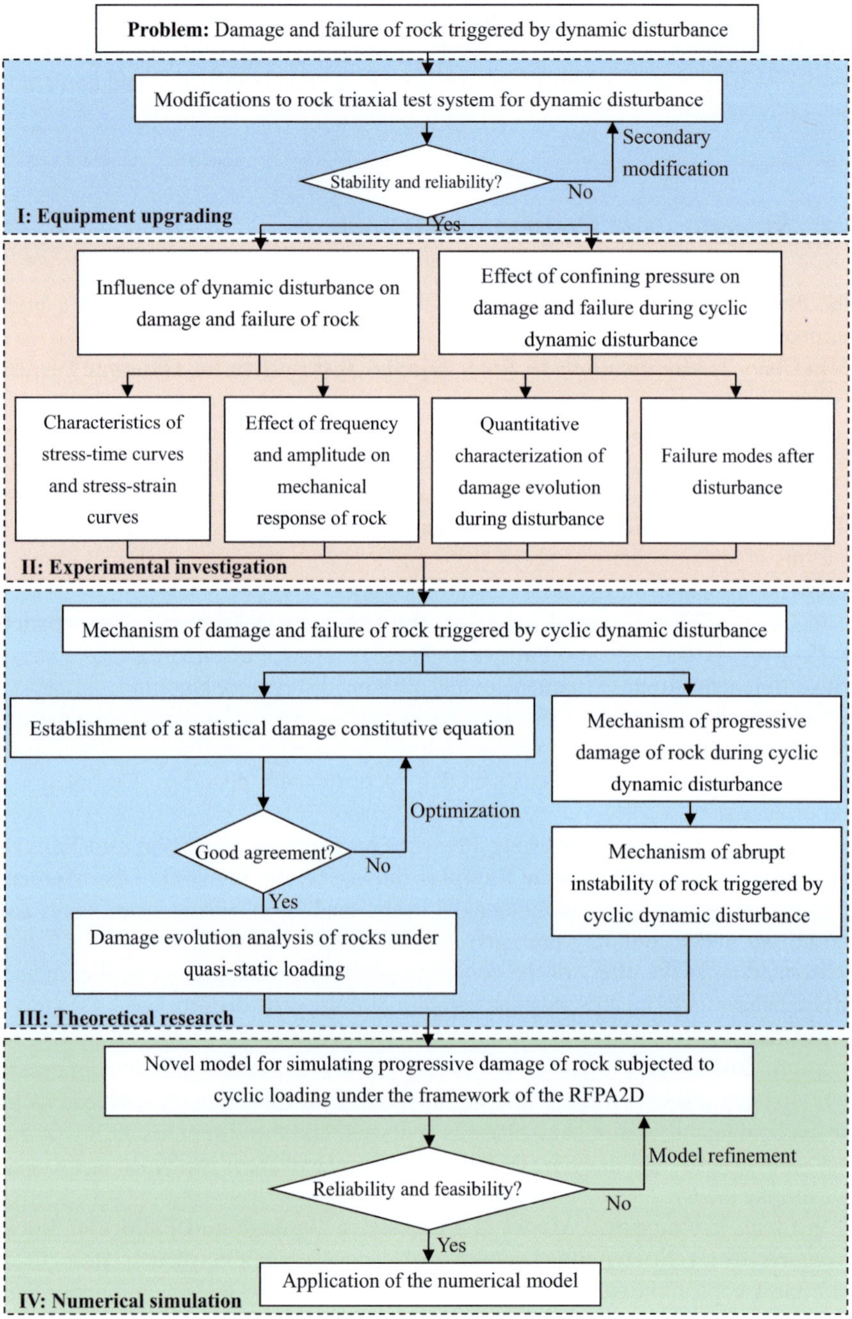

Fig. 1.5 Flow chart of this research

In Chap. 8, **Concluding Remarks and Prospects**: Conclusions are drawn for damage and failure of rock triggered by cyclic dynamic disturbance. The suggestions for further research work are presented.

References

Attewell PB, Farmer IW (1973) Fatigue behaviour of rock. Int J Rock Mech Min Sci Geomech Abstr 10(1):1–9. https://doi.org/10.1016/0148-9062(73)90055-7

Bagde MN, Petroš V (2005a) Fatigue properties of intact sandstone samples subjected to dynamic uniaxial cyclical loading. Int J Rock Mech Min Sci 42(2):237–250. https://doi.org/10.1016/j.ijrmms.2004.08.008

Bagde MN, Petroš V (2005b) Waveform effect on fatigue properties of intact sandstone in uniaxial cyclical loading. Rock Mech Rock Eng 38(3):169–196. https://doi.org/10.1007/s00603-005-0045-8

Beom HG, Zhuo XR, Cui CB (2013) Tunneling cracks in the adhesive layer of an orthotropic sandwich structure. Int J Eng Sci 63:40–51. https://doi.org/10.1016/j.ijengsci.2012.11.001

Bhattacharya D, Ghosh JK, Samadhiya NK (2012) Review of geohazard warning systems toward development of a popular usage geohazard warning communication system. Nat Hazard Rev 13(4):260–271. https://doi.org/10.1061/(ASCE)NH.1527-6996.0000078

Chen ZY, Wang Z, Xi H et al (2016) Recent advances in high slope reinforcement in China: case studies. J Rock Mech Geotech Eng 8(6):775–788. https://doi.org/10.1016/j.jrmge.2016.11.001

Chen CF, Xu T, Heap MJ et al (2018) Influence of unloading and loading stress cycles on the creep behavior of Darley Dale Sandstone. Int J Rock Mech Min Sci 112:55–63. https://doi.org/10.1016/j.ijrmms.2018.09.002

Deng HF, Hu Y, Li JL et al (2017) Effects of frequency and amplitude of cyclic loading on the dynamic characteristics of sandstone. Rock Soil Mech 38(12):3402–3409 (in Chinese). https://doi.org/10.16285/j.rsm.2017.12.002

Du K, Li XB, Ma CD (2014) Development of rock true triaxial dynamic disturbance test system and its application. Exp Technol Manag 31(12):35–40 (in Chinese). https://doi.org/10.16791/j.cnki.sjg.2014.12.010

Du K, Li XB, Li DY et al (2015) Failure properties of rocks in true triaxial unloading compressive test. Trans Nonferrous Metals Soc China 25(2):571–581. https://doi.org/10.1016/S1003-6326(15)63639-1

Du K, Tao M, Li XB et al (2016) Experimental study of slabbing and rockburst induced by true-triaxial unloading and local dynamic disturbance. Rock Mech Rock Eng 49(6):3437–3453. https://doi.org/10.1007/s00603-016-0990-4

Feng GL, Feng XT, Chen BR et al (2019) Effects of structural planes on the microseismicity associated with rockburst development processes in deep tunnels of the Jinping-II Hydropower Station, China. Tunn Undergr Space Technol 84:273–280. https://doi.org/10.1016/j.tust.2018.11.008

Fu B, Hu LH, Tang CA (2020) Experimental and numerical investigations on crack development and mechanical behavior of marble under uniaxial cyclic loading compression. Int J Rock Mech Min Sci 130:104289. https://doi.org/10.1016/j.ijrmms.2020.104289

Ge XR, Jiang Y, Lu YD (2003) Testing study on fatigue deformation law of rock under cyclic loading. Chin J Rock Mech Eng 22(10):1581–1585 (in Chinese)

Ghassemi A (2012) A review of some rock mechanics issues in geothermal reservoir development. Geotech Geol Eng 30(3):647–664. https://doi.org/10.1007/s10706-012-9508-3

Gong FQ, Li XB, Liu XL et al (2010) Experimental study of dynamic characteristics of sandstone under one-dimensional coupled static and dynamic loads. Chin J Rock Mech Eng 29(10):2076–2085 (in Chinese)

Gong FQ, Li XB, Liu XL (2011) Preliminary experimental study of characteristics of rock subjected to 3D coupled static and dynamic loads. Chin J Rock Mech Eng 30(6):1179–1190 (in Chinese)

Home L (2016) Hard rock TBM tunneling in challenging ground: Developments and lessons learned from the field. Tunn Undergr Space Technol 57:27–32. https://doi.org/10.1016/j.tust.2016.01.008

Hu LH, Ma K, Liang X et al (2018) Experimental and numerical study on rockburst triggered by tangential weak cyclic dynamic disturbance under true triaxial conditions. Tunn Undergr Space Technol 81:602–618. https://doi.org/10.1016/j.tust.2018.08.014

Hu LH, Ju MH, Zhao P et al (2023) Deformation characteristics and novel strain criteria of strainbursts induced by low-frequency cyclic disturbance. Deep Under Sci Eng 2(1):52–60. https://doi.org/10.1002/dug2.12029

Huang X, Liu QS, Liu B et al (2017) Experimental study on the dilatancy and fracturing behavior of soft rock under unloading conditions. Int J Civ Eng 15:921–948. https://doi.org/10.1007/s40999-016-0144-9

Hudson JA, Stephansson O, Andersson J et al (2001) Coupled T–H–M issues relating to radioactive waste repository design and performance. Int J Rock Mech Min Sci 38(1):143–161. https://doi.org/10.1016/S1365-1609(00)00070-8

Hunsche U, Hampel A (1999) Rock salt—the mechanical properties of the host rock material for a radioactive waste repository. Eng Geol 52(3–4):271–291. https://doi.org/10.1016/S0013-7952(99)00011-3

Izadi G, Elsworth D (2015) The influence of thermal-hydraulic-mechanical- and chemical effects on the evolution of permeability, seismicity and heat production in geothermal reservoirs. Geothermics 53:385–395. https://doi.org/10.1016/j.geothermics.2014.08.005

Jiang Y, Ge XR, Ren JX (2004) Deformation rules and acoustic emission characteristics of rocks in process of fatigue failure. Chin J Rock Mech Eng 23(11):1810–1814 (in Chinese)

Jiang D, Xie K, Chen J et al (2019) Experimental analysis of sandstone under uniaxial cyclic loading through acoustic emission statistics. Pure Appl Geophys 176:265–277. https://doi.org/10.1007/s00024-018-1960-4

Jin JS, Shi ZQ, Fang H (1992) The effect of cyclic loading-unloading on the unstable rupture of rock samples. Chin J Rock Mech Eng 11(4):384–389 (in Chinese)

Li N, Chen W, Zhang P et al (2001) The mechanical properties and a fatigue-damage model for jointed rock masses subjected to dynamic cyclical loading. Int J Rock Mech Min Sci 38(7):1071–1079. https://doi.org/10.1016/S1365-1609(01)00058-2

Li XB, Zhou ZL, Lok TS et al (2008a) Innovative testing technique of rock subjected to coupled static and dynamic loads. Int J Rock Mech Min Sci 45(5):739–748. https://doi.org/10.1016/j.ijrmms.2007.08.013

Li XB, Zhou ZL, Ye YZ et al (2008b) Study of rock mechanical characteristics under coupled static and dynamic loads. Chin J Rock Mech Eng 27(7):1387–1395 (in Chinese). https://doi.org/10.3321/j.issn:1000-6915.2008.07.011

Li XB, Zhou ZL, Zhao FJ et al (2009) Mechanical properties of rock under coupled static-dynamic loads. J Rock Mech Geotech Eng 1(1):41–47. https://doi.org/10.3724/SP.J.1235.2009.00041

Li XB, Du K, Li DY (2015a) True triaxial strength and failure modes of cubic rock specimens with unloading the minor principal stress. Rock Mech Rock Eng 48(6):2185–2196. https://doi.org/10.1007/s00603-014-0701-y

Li XB, Weng L, Xie XF et al (2015b) Study on the degradation of hard rock with a pre-existing opening under static-dynamic loadings using nuclear magnetic resonance technique. Chin J Rock Mech Eng 34(10):1985–1993 (in Chinese). https://doi.org/10.13722/j.cnki.jrme.2015.1026

Li XB, Gong FQ, Tao M et al (2017) Failure mechanism and coupled static-dynamic loading theory in deep hard rock mining: a review. J Rock Mech Geotech Eng 9:767–782. https://doi.org/10.1016/j.jrmge.2017.04.004

Li DY, Xiao P, Han ZY et al (2020) Mechanical and failure properties of rocks with a cavity under coupled static and dynamic loads. Eng Fract Mech 225:106195. https://doi.org/10.1016/j.engfracmech.2018.10.021

Lin TJ, Wu XM (1976) Deformation characteristics of rock. Metal Mine 4:21–29 (in Chinese)
Lin ZY, Wu YS (1987) Strength and deformability of rock under cyclic loading. Rock Soil Mech 8(3):33–39 (in Chinese). https://doi.org/10.16285/j.rsm.1987.03.008
Liu EL, He SM (2012) Effects of cyclic dynamic loading on the mechanical properties of intact rock samples under confining pressure conditions. Eng Geol 125(27):81–91. https://doi.org/10.1016/j.enggeo.2011.11.007
Liu JF, Xu J, Li QS et al (2010) Experimental research on damping parameters of rock under cyclic loading. Chin J Rock Mech Eng 29(5):1036–1041 (in Chinese)
Liu EL, Huang RQ, He SM (2012a) Effects of frequency on the dynamic properties of intact rock samples subjected to cyclic loading under confining pressure conditions. Rock Mech Rock Eng 45(1):89–102. https://doi.org/10.1007/s00603-011-0185-y
Liu JF, Xie HP, Xu J et al (2012b) Discussion on deformation and damping parameters of rock under cyclic loading. Chin J Rock Mech Eng 31(4):770–777 (in Chinese). https://doi.org/10.3969/j.issn.1000-6915.2012.04.016
Liu SH, Qin ZH, Lou JF et al (2014) Experimental study of dynamic failure characteristics of coal-rock compound under one-dimensional static and dynamic loads. Chin J Rock Mech Eng 33(10):2064–2075 (in Chinese). https://doi.org/10.13722/j.cnki.jrme.2014.10.013
Meng QB, Zhang MW, Han LJ et al (2016) Effects of acoustic emission and energy evolution of rock specimens under the uniaxial cyclic loading and unloading compression. Rock Mech Rock Eng 49:3873–3886. https://doi.org/10.1007/s00603-016-1077-y
Meng QB, Zhang MW, Han LJ et al (2018) Acoustic emission characteristics of red sandstone specimens under uniaxial cyclic loading and unloading compression. Rock Mech Rock Eng 51:969–988. https://doi.org/10.1007/s00603-017-1389-6
Ni XH, Zhu ZD, Li XJ et al (2011) Quantitative test study of meso-damage of rock under cyclic load. Rock Soil Mech 32(7):1991–1995 (in Chinese). https://doi.org/10.3969/j.issn.1000-7598.2011.07.012
Pruess K (2006) Enhanced Geothermal Systems (EGS) using CO_2, as working fluid—a novel approach for generating renewable energy with simultaneous sequestration of carbon. Geothermics 35(4):351–367. https://doi.org/10.1016/j.geothermics.2006.08.002
Qian QH (2012) Challenges faced by underground projects construction safety and countermeasures. Chin J Rock Mech Eng 31(10):1945–1956 (in Chinese). https://doi.org/10.3969/j.issn.1000-6915.2012.10.001
Shahnazari H, Dehnavi Y, Alavi AH (2010) Numerical modeling of stress-strain behavior of sand under cyclic loading. Eng Geol 116(1–2):53–72. https://doi.org/10.1016/j.enggeo.2010.07.007
Singh SK (1989) Fatigue and strain hardening behaviour of graywacke from the flagstaff formation, new south Wales. Eng Geol 26(2):171–179. https://doi.org/10.1016/0013-7952(89)90005-7
Song HP, Zhang H, Kang YL et al (2013) Damage evolution study of sandstone by cyclic uniaxial test and digital image correlation. Tectonophysics 608:1343–1348. https://doi.org/10.1016/j.tecto.2013.06.007
Su GS, Feng XT, Wang JH et al (2017) Experimental study of remotely triggered rockburst induced by a tunnel axial dynamic disturbance under true-triaxial conditions. Rock Mech Rock Eng 50(4):2207–2226. https://doi.org/10.1007/s00603-017-1218-y
Sun LD, Fang CL, Sa LM et al (2015) Innovation and prospect of geophysical technology in the exploration of deep oil and gas. Pet Explor Dev 42(4):454–465. https://doi.org/10.1016/S1876-3804(15)30038-0
Taheri A, Royle A, Yang Z et al (2016) Study on variations of peak strength of a sandstone during cyclic loading. Geomech Geophys Geo-Energy Geo-Resour 2(1):1–10. https://doi.org/10.1007/s40948-015-0017-8
Tang CA, Li LC, Xu NW et al (2015) Microseismic monitoring and numerical simulation on the stability of high-steep rock slopes in hydropower engineering. J Rock Mech Geotech Eng 7(5):493–508. https://doi.org/10.1016/j.jrmge.2015.06.010

Tao ZY, Mo HH (1990) An experimental study and analysis of the behaviour of rock under cyclic loading. Int J Rock Mech Min Sci Geomech Abstr 27(1):51–56. https://doi.org/10.1016/0148-9062(90)90008-P

Tao M, Ma A, Cao WZ et al (2017) Dynamic response of pre-stressed rock with a circular cavity subject to transient loading. Int J Rock Mech Min Sci 99:1–8. https://doi.org/10.1016/j.ijrmms.2017.09.003

Taylor LM, Chen EP, Kuszmaul JS (1986) Microcrack-induced damage accumulation in brittle rock under dynamic loading. Comput Methods Appl Mech Eng 55(3):301–320. https://doi.org/10.1016/0045-7825(86)90057-5

Tsang CF, Barnichon JD, Birkholzer J (2012) Coupled thermo-hydro-mechanical processes in the near field of a high-level radioactive waste repository in clay formations. Int J Rock Mech Min Sci 49(1):31–44. https://doi.org/10.1016/j.ijrmms.2011.09.015

Walton G, Arzúa J, Alejano LR et al (2015) A laboratory-testing-based study on the strength, deformability, and dilatancy of carbonate rocks at low confinement. Rock Mech Rock Eng 48:941–958. https://doi.org/10.1007/s00603-014-0631-8

Wang RH, Li JL, Jiang YZ et al (2010) Experimental research on influence of cyclic loading and unloading on rock mass residual strength. Chin J Rock Mech Eng 29(10):2103–2109 (in Chinese)

Wang ZC, Li SC, Qiao LP et al (2013) Fatigue behavior of granite subjected to cyclic loading under triaxial compression condition. Rock Mech Rock Eng 46(6):1603–1615. https://doi.org/10.1007/s00603-013-0387-6

Wang J, Chen L, Su R et al (2018) The Beishan underground research laboratory for geological disposal of high-level radioactive waste in China: planning, site selection, site characterization and in situ tests. J Rock Mech Geotech Eng 10(3):411–435. https://doi.org/10.1016/j.jrmge.2018.03.002

Xia D, Yang TH, Wang PT et al (2014) Analysis on damage and energy in deformation and fracture of saturated rock subjected to cyclic loading and unloading. J Northeast Univ (Nat Sci) 35(6):867–870 (in Chinese). https://doi.org/10.3969/j.issn.1005-3026.2014.06.024

Xiao JQ, Ding DX, Jiang FL et al (2010) Fatigue damage variable and evolution of rock subjected to cyclic loading. Int J Rock Mech Min Sci 47(3):461–468. https://doi.org/10.1016/j.ijrmms.2009.11.003

Yang YJ, Song Y, Chu J (2007) Experimental study on characteristics of strength and deformation of coal under cyclic loading. Chin J Rock Mech Eng 26(1):201–205 (in Chinese)

Yang SQ, Tian WL, Ranjith PG (2017) Experimental investigation on deformation failure characteristics of crystalline marble under triaxial cyclic loading. Rock Mech Rock Eng 50(1):1–19. https://doi.org/10.1007/s00603-017-1262-7

Zamani Z, Rahimpour-Bonab H, Littke R (2023) Coal petrology, sedimentology and depositional environment of the Parvadeh coals in the Upper Triassic, Tabas Block of Central-East Iran. Int J Coal Sci Technol 10:40. https://doi.org/10.1007/s40789-023-00600-w

Zhang HH, Yan YD, Yu HZ et al (2004) Acoustic emission experimental research on large-scaled rock failure under cyclic loading—fracture precursor of rock. Chin J Rock Mech Eng 23(21):3621–3628 (in Chinese)

Zhang QX, Ge XR, Huang M et al (2006) Testing study on fatigue deformation law of red-sandstone under triaxial compression with cyclic loading. Chin J Rock Mech Eng 25(3):473–478 (in Chinese)

Zhang H, Song HP, Kang YL et al (2012) Experimental analysis on deformation evolution and crack propagation of rock under cyclic indentation. Rock Mech Rock Eng 46:1053–1059. https://doi.org/10.1007/s00603-012-0309-z

Zhang CQ, Feng XT, Zhou H et al (2013) Rockmass damage development following two extremely intense rockbursts in deep tunnels at Jinping II hydropower station, southwestern China. Bull Eng Geol Environ 72:237–247. https://doi.org/10.1007/s10064-013-0470-y

Zhang SS, Liu EL, Zhang JH (2014) Experimental study of fatigue damage properties of sandstone samples under cyclic loading with low frequencies. Chin J Rock Mech Eng 33(S1):3212–3218 (in Chinese). https://doi.org/10.13722/j.cnki.jrme.2014.s1.087

Zhao XG, Wang J, Cai M et al (2014) Influence of unloading rate on the strainburst characteristics of Beishan granite under true-triaxial unloading conditions. Rock Mech Rock Eng 47:467–483. https://doi.org/10.1007/s00603-013-0443-2

Zhao YL, Zhang LY, Wang WJ et al (2017) Creep behavior of intact and cracked limestone under multi-level loading and unloading cycles. Rock Mech Rock Eng 50:1409–1424. https://doi.org/10.1007/s00603-017-1187-1

Zhou ZL (2007) Study on experimental and mechanical behaviors of rock under static-dynamic coupling loads. Central South University (in Chinese)

Zhou H, Pan PZ, Feng XT (2006) Plane elastoplastic cellular automata model of failure process of rocks under cyclic loading. Chin J Rock Mech Eng 25(S2):3623–3628 (in Chinese). https://doi.org/10.3321/j.issn:1000-6915.2006.z2.045

Zhou ZL, Li XB, Zou Y et al (2014) Dynamic Brazilian tests of granite under coupled static and dynamic loads. Rock Mech Rock Eng 47(2):495–505. https://doi.org/10.1007/s00603-013-0441-4

Zhou L, Zhu ZM, Wang M et al (2018) Dynamic propagation behavior of cracks emanating from tunnel edges under impact loads. Soil Dyn Earthq Eng 105:119–126. https://doi.org/10.1016/j.soildyn.2017.12.012

Zhou HW, Wang ZH, Wang CS et al (2019) On acoustic emission and post-peak energy evolution in Beishan granite under cyclic loading. Rock Mech Rock Eng 52:283–288. https://doi.org/10.1007/s00603-018-1614-y

Zhu ML, Zhu ZD, Li G et al (2009) Experimental study of dynamic characteristics of granite under cyclic loading. Chin J Rock Mech Eng 28(12):2520–2526 (in Chinese). https://doi.org/10.3321/j.issn:1000-6915.2009.12.019

Zhu WC, Li ZH, Zhu L et al (2010) Numerical simulation on rockburst of underground opening triggered by dynamic disturbance. Tunn Undergr Space Technol 25(5):587–599. https://doi.org/10.1016/j.tust.2010.04.004

Zhu WC, Bai Y, Li XB et al (2012) Numerical simulation on rock failure under combined static and dynamic loading during SHPB tests. Int J Impact Eng 49(2):142–157. https://doi.org/10.1016/j.ijimpeng.2012.04.002

Zuo YJ (2005) Study on failure and fragmentation characteristics of rock under static-dynamic coupling loading. Central South University of Technology (in Chinese)

Zuo YJ, Li XB, Zhou ZL et al (2005) Damage and failure rule of rock undergoing uniaxial compressive load and dynamic load. J Cent South Univ Technol 12(6):742–748. https://doi.org/10.1007/s11771-005-0080-3

Zuo YJ, Li XB, Tang CA et al (2006) Experimental investigation on failure of rock subjected to 2D dynamic-static coupling loading. Chin J Rock Mech Eng 25(9):1809–1809 (in Chinese). https://doi.org/10.3321/j.issn:1000-6915.2006.09.012

Chapter 2
Modifications to Rock Triaxial Test System for Dynamic Disturbance

2.1 Introduction

From the current development of the energy industry, mineral resources in shallow parts have gradually become dried up, while renewable energy and clean energy are developing rapidly, making the development and utilization of mineral resources in the deep parts of the earth become more and more common. From the current development of traffic engineering, urban ground transportation is gradually becoming saturated, and the development of urban underground space and even deep underground is in full swing. It means that people need to normalize the development of deep mineral resources and rationalize the utilization of urban underground space. Due to the coupling condition of high ground stress and high temperature of deep rock mass in general, the deep excavation of rock mass and the design and construction of deep roadway surrounding rock are quite different from shallow engineering. At the same time, it is not reasonable to use "monotonic loading" to simulate the mechanical properties of deep rocks in the laboratory: engineering rock mass is not in a monotonic loading state under ideal conditions such as constant speed and constant confining pressure, but forms a free surface with a complex and changeable unloading process accompanied by cyclic dynamic effects, such as blasting vibration, rock breaking with disc cutters of shield machine, rockburst of rock mass in high-stress area, etc., while enduring high ground stress in depth (Liu et al. 2018; Naji et al. 2018). Rock mass is often in the process of coupling loading and unloading with dynamic disturbance (He et al. 2023a). At present, although scholars and engineers have gradually turned their attention to deep engineering, the experience in design and construction is still not rich enough, the understanding of the mechanical response of deep rock mass under the action of multiple loads, such as cyclic loading–unloading and dynamic disturbance, is still insufficient (He et al. 2023b). The ability to control rock engineering stability and deformation, and the ability to predict and prevent deep rock engineering diseases and accidents are still lacking (Yang et al. 2023). Therefore, it is of great practical significance and necessity to

X. Chen, *Damage and Failure of Rock Triggered by Dynamic Disturbance*,
https://doi.org/10.1007/978-981-96-9486-0_2

develop a rock triaxial test system that is more in line with the actual stress state of deep rock mass and perform loading–unloading and dynamic disturbance at the same time, and to explore the corresponding mechanical response of deep rock mass.

In recent years, many scholars have made a great amount of exploration on the dynamic disturbance of rocks, including the strength characteristics of rocks under dynamic disturbance (Ma et al. 2013) and the internal mechanism of deformation and failure (Fuenkajorn and Phueakphum 2010; Zhu et al. 2010). To a certain extent, the fatigue properties under cyclic loading can also belong to the category of dynamic disturbance, and some scholars have done corresponding research in this area (Erarslan and Williams 2012). Some other have also explained the mechanical response of rocks under dynamic disturbance from the perspectives of hysteresis effect (Azra et al. 1998), damping parameters (Liu et al. 2010), and energy absorption and dissipation (Song et al. 2012). In addition, Li et al. (2007), Zuo et al. (2006), Zhou et al. (2014), Gong et al. (2014) also discovered the characteristics of damage and failure of rocks under high stress levels induced by dynamic disturbance through experimental means, and tried to explain its essential mechanism from damage variables, nonlinear damage models, etc.

It has to be said that previous research has deepened the understanding of the mechanical response of rock under the action of dynamic disturbance and its inherent damage and failure mechanism. However, there is still a problem that has not yet reached a consensus in their research: the dynamic disturbance of rocks and the high stress environment they are in actually act synergistically, and the coupling effects result in deformation, damage, and instability of rocks, but most previous studies have just studied the effects of the two in isolation without taking into account their synergy, or have taken into account their synergy but ignored the frequency, amplitude and other influencing factors of the dynamic disturbance. Therefore, the disturbance of rocks by external forces under conditions of high stress is one of the hot issues currently widely studied in the field of rock mechanics.

In view of this, based on a conventional electro-hydraulic servo rock triaxial testing system (TFD-2000/D) produced by Changchun Keyi Testing Instrument Co., Ltd., a rock dynamic disturbance servo triaxial test system is independently developed, which can independently carry out dynamic disturbance closed-loop control, confining pressure closed-loop control and axial pressure closed-loop control, and realize the dynamic disturbance and pre-static load synergistic action on rock samples. In this chapter, the structure of the test system and its functions are described in detail. Section 2.2 mainly deals with the preliminary development of the test system. Although the dynamic disturbance triaxial test system can meet the basic functional requirements, there are some shortcomings in sample installation and sealing ability. Accordingly, in Sect. 2.3, the system is reconstructed for the second time in view of these shortcomings in the preliminary design. Then, the limitations of the modified equipment and the stability and reliability of the equipment are introduced in Sects. 2.4 and 2.5, respectively.

2.2 Preliminary Modification to Rock Triaxial Test System

In the early days of rock mechanics research, the basic properties of rocks, such as elastic modulus, uniaxial or triaxial compressive strength, Poisson's ratio, etc., were mainly completed through ordinary universal material testing machines. As people gradually understand and master the principles and technologies of rigid testing machines, special rock triaxial testing systems have begun to be used to test the basic mechanical properties of rocks and concrete (Jaeger 1967; Goodman 1989; Paterson and Wong 2005). Up to now, conventional rock electro-hydraulic servo triaxial testing equipment has become one of the most common, mature and stable equipment in many rock mechanics laboratories. The TFD-2000/D rock electro-hydraulic servo triaxial testing system designed and produced by Changchun Keyi Testing Instrument Co., Ltd. is such an apparatus that is capable of controlling and loading of axial stress, confining pressure, and pore pressure independently. It is usually used to carry out uniaxial compression, triaxial compression, cyclic loading and unloading, permeability testing, creep loading and other mechanical tests on rock materials. However, the combined effects of the dynamic disturbance and triaxial stresses can hardly be simulated in laboratory because these loads are difficult to impose simultaneously on the rock sample with a traditional rock triaxial test system. Therefore, a servo-controlled rock triaxial test system with the function of imposing external disturbance load is newly developed.

Figure 2.1 shows the physical picture and structural schematic diagram of the initially developed dynamic disturbance electro-hydraulic servo rock triaxial testing system. The part circled by the red rectangular box in Fig. 2.1a is the developed dynamic disturbance loading device. It can be seen from the overall testing machine that although the dynamic disturbance loading device shares the trolley guide rail and the loading frame with the confining pressure chamber, its base is connected to another separate trolley, forming a separate triaxial loading device. When performing a dynamic disturbance triaxial test, we can loosen the bolts on the upper part of the metal confining cylinder used in conventional triaxial tests, lower the confining cylinder, and push it out together with the connected base trolley. Then push the trolley used by the dynamic disturbance loading device into the main loading frame, use the bottom jack to lift the trolley to the lower edge of the positioning frame of the disturbance device, and tighten the bolts. Similarly, if conventional triaxial testing is required, the dynamic disturbance loading device is removed and replaced with a conventional confining pressure chamber and other related devices. Figure 2.1b shows the schematic diagram of the front elevation of a separate dynamic disturbance triaxial test device.

The developed dynamic disturbance rock servo triaxial test system can independently carry out the loading and unloading of axial pressure, confining pressure, dynamic perturbation, precise control as well as data acquisition. These four subsystems, i.e., axial loading subsystem, confining pressure subsystem, dynamic disturbance subsystem, and data collection and recording subsystem, are introduced in detail below.

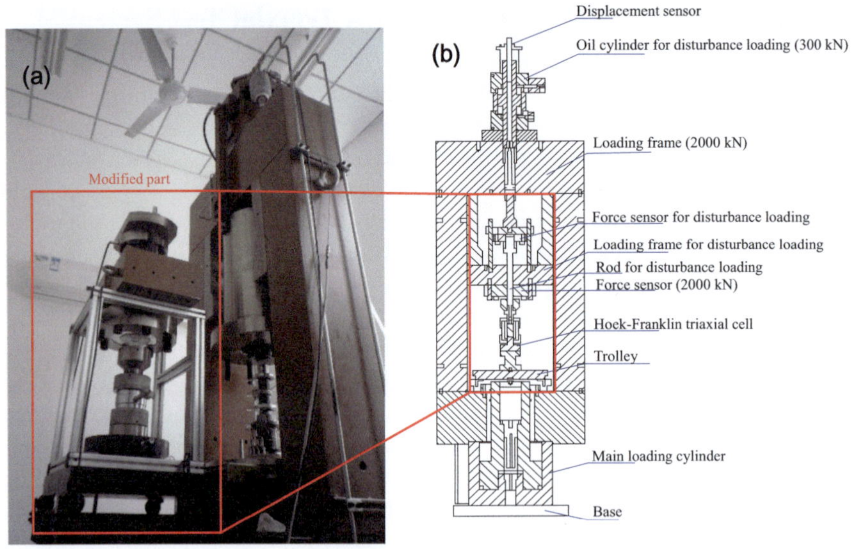

Fig. 2.1 Preliminary modification to servo-controlled rock triaxial test system: **a** physical photograph, **b** schematic diagram

2.2.1 Axial Loading Subsystem

The developed dynamic disturbance rock servo triaxial test system shares a same axial loading subsystem with the original conventional triaxial test system when applying axial load. The main components of the axial loading subsystem include servo oil source, axial EDC controller, and electronic computer. They are connected to the load sensor installed on the top platen of the testing machine. Through the data collection and feedback of the load sensor, PID (proportion, integral, differential) closed-loop control is realized to accurately adjust the amount of oil in and out of the oil cylinder at the bottom of the main testing machine, thereby lifting the base trolley and controlling the loading, unloading and stable maintenance of axial loading. The EDC controller and PID closed loop are capable of controlling the axial load with high accuracy, reaching $\pm$ 0.01 kN. The maximum axial load that can be applied is up to 2000 kN. The large axial load application capacity ensures that the test system can not only carry out triaxial compression tests on lower-strength rocks, but also meet the axial load application capacity on high-strength rocks, concrete and other samples, and obtain their basic data such as strength and complete stress–strain curves.

2.2.2 Confining Pressure Subsystem

In order to independently apply and control the confining pressure, The Hoek–Franklin triaxial pressure chamber, which has some advantages of small chamber, simple structure, easy operation, and low price (Hoek 1968; Franklin and Hoeck 1970), is mounted on the developed dynamic disturbance rock servo triaxial test system. Except for replacing the self-balancing pressure chamber of the conventional triaxial testing machine with a Hoek–Franklin triaxial pressure chamber, other confining pressure loading components, including the confining pressure flushing liquid cylinder, the air pump used for air drive pressurization, and EDC controller and electronic computer for closed loop control, have not been changed in the modification. When applying confining pressure, a gas drive pump is first used to drive the pressure to a value near the target value that can be read from the pressure gauge on the panel. Then the gas drive pump valve is closed and the target confining pressure is sent to the controller through the computer. The controller begins to further adjust the confining pressure and at the same time feeds back the value collected by the pressure sensor to the computer. The computer will also send the target confining pressure value to the controller based on the feedback signal, and in this cycle, the confining pressure is loaded and unloaded, and finally maintained at the target. This process makes the confining pressure subsystem achieve stable and accurate closed-loop control and application of confining pressure. The confining pressure subsystem is capable of providing a maximum confining pressure of up to 70 MPa with accuracy of $\pm$ 0.01 MPa.

Figure 2.2 gives the schematic diagram and physical photograph of the Hoek–Franklin triaxial pressure chamber. As shown in Fig. 2.2a, in order to realize the dynamic disturbance function, a disturbance rod, which is one of the main components of this modification and was not available on the previous Hoek–Franklin triaxial pressure chamber, was added to the upper platen of the chamber for the conduction and application of dynamic disturbance loads. The disturbance rod transfers axial disturbance loads to the rock sample under the control of the dynamic disturbance loading subsystem which will detailed in Sect. 2.2.3.

2.2.3 Dynamic Disturbance Loading Subsystem

Dynamic disturbance loading subsystem, the components of which include disturbance cylinder, servo oil source, servo controller, etc., is a set of loading and control subsystem parallel to axial pressure and confining pressure. An EDC580 all-digital servo controller (Fig. 2.3) manufactured by DOLI Elektronik GmbH, Germany is configured as the control terminal, and it is capable of accurate measurement and control of load and deformation. The built-in self-protection program can also largely avoid damage to load sensors and deformation sensors in the tests. A dedicated software equipped with the EDC 580 servo controller is DOLI Installation Center

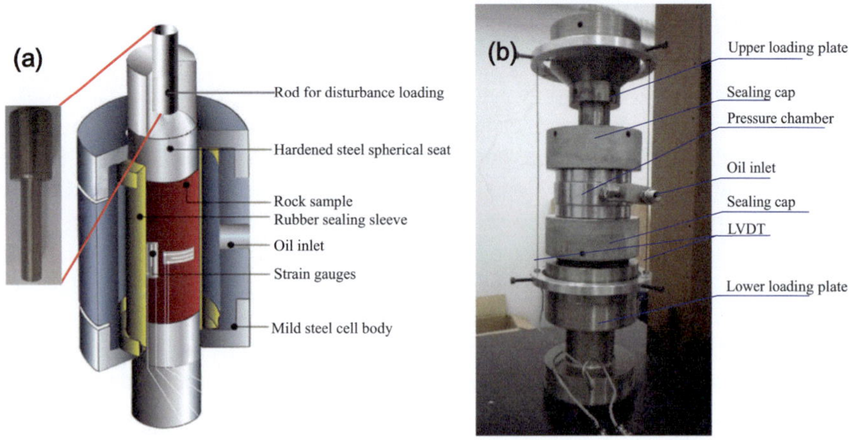

Fig. 2.2 The Hoek–Franklin triaxial pressure chamber. **a** Schematic diagram, **b** physical photograph

(V3.3.1.0). The test parameters required when dynamic disturbance is applied, such as amplitude, frequency, phase, etc., are completed by filling in the menu of the software, which is easy to operate and convenient. In addition, the software also has the functions of self-diagnosis and automatic calibration.

When the dynamic disturbance is applied, the computer sends the dynamic disturbance parameters, and the servo controller controls the oil cylinder on the top of the machine to act on the rock sample through the closed-loop control. The frequency of the applied dynamic disturbance ranges from 0 to 70 Hz, and the maximum disturbance load that can be applied is up to 300 kN. As shown in Fig. 2.4, the waveforms that the dynamic disturbance loading subsystem is capable of generating includes sinusoidal wave, ramp wave, ideal and actual square wave for cyclic disturbance load. The amplitude control modes of the waveforms can be divided into displacement

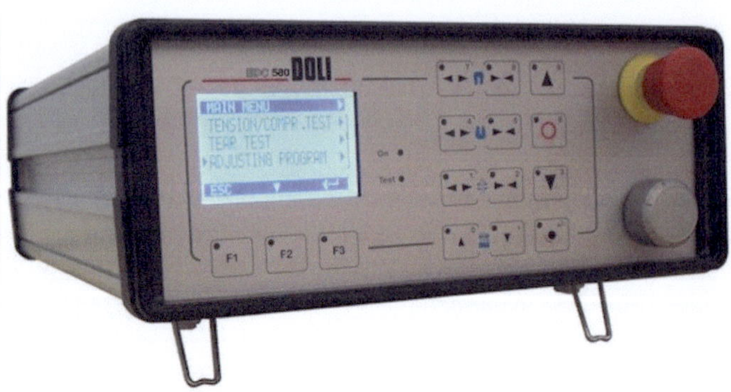

Fig. 2.3 EDC 580 all-digital servo controller

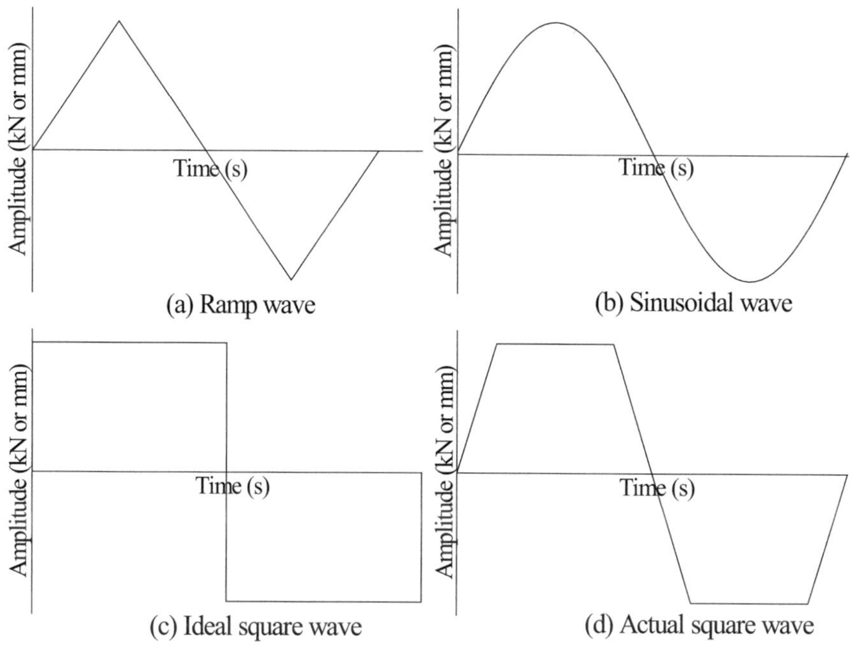

Fig. 2.4 Schematic diagrams of four different waveforms of disturbance that the modified servo-controlled rock triaxial test system enable to generate

control and load control. Figure 2.5 shows the control and application of sinusoidal wave disturbance of three different frequencies and amplitudes in the actual test. As can be seen from the figure, even if the applied waveform frequency reaches its maximum (70 Hz), the command signal and the feedback signal are still relatively consistent, indicating that the subsystem achieves a good effect of application and control of dynamic disturbance.

It should be noted that as can be seen from Fig. 2.2a, the end face of the disturbance rod placed in the upper steel platen is smaller than the end face of the rock sample. Therefore, when the dynamic disturbance is applied on the rock sample using the preliminarily developed device, the dynamic disturbance will transfer in the form of point disturbance in the axial direction.

2.2.4 *Data Collection and Recording Subsystem*

For dynamic disturbance test, the mechanical response of rock in each disturbance period needs to be recorded in real time, which has the characteristics of large amount

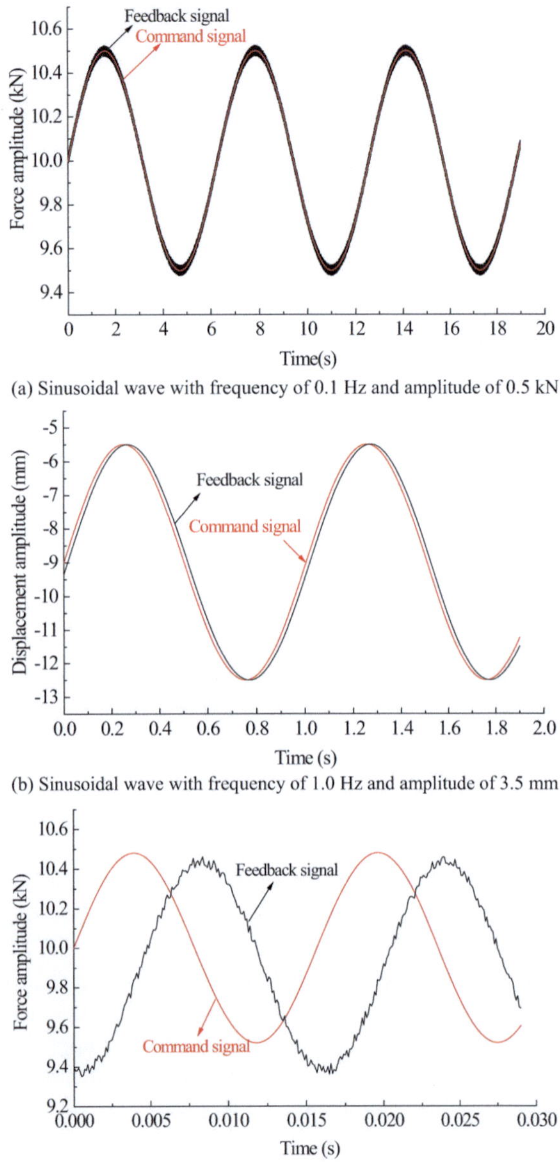

(a) Sinusoidal wave with frequency of 0.1 Hz and amplitude of 0.5 kN

(b) Sinusoidal wave with frequency of 1.0 Hz and amplitude of 3.5 mm

(c) Sinusoidal wave with frequency of 70 Hz and amplitude of 0.5 kN

Fig. 2.5 Three kinds of dynamic disturbance curves with different frequencies and amplitudes in actual tests

of data, so the automatic data collection, recording and storage is particularly important. The data acquisition, record storage and other functions of the dynamic disturbance triaxial testing machine are mainly completed through a dedicated software, which is modified on the basis of the original test and record software developed by Changchun Keyi Experimental Instrument Co., LTD. In view of the large amount of dynamic disturbance test data, the software has been specially modified and compiled to have capacity of collecting and recording 300,000 lines of data at one time, so as to efficiently and accurately record key information such as strain, confining pressure, axial stress and test duration during high frequency dynamic disturbance.

2.3 Secondary Modification to the Rock Triaxial Test System

Although the preliminary modified rock triaxial test system can meet the basic requirement of applying external axial dynamic disturbance load during triaxial compression, it still has some defects and limitations as follow:

(1) In the preliminary triaxial test system for dynamic disturbed rock, data on axial and radial deformation must be collected by means of strain gauges attached to the surface of the rock sample. The strain gauge and the wires attached to it must be inserted into a special rubber sealing sleeve for isolating the confining oil from the rock sample (as shown in Fig. 2.2a), and then the rubber sleeve and rock sample are placed in the Hoek–Franklin triaxial pressure chamber. When the rock sample with strain gauge is inserted into the rubber sleeve, it is very easy to cause wire damage and strain gauges to fall off, resulting in failure to collect strain data and apply load.

(2) Whether the strain gauges are used to measure axial strain or circumferential strain, the measured data can only reflect the deformation of the rock surface near the location of the strain gauge. However, strain localization often occurs in the process of rock triaxial test, such as local shear, swelling, splitting, etc., leading to imprecise results in the data recorded due to the different position that the gauges paste on and the lower measuring accuracy of the bounded gauges than circumferential extensometers. In the preliminary modification, although a set of external axial extensometer (LVDT) (as shown in Fig. 2.2b) equipped for measuring axial strain improves the axial strain measurement accuracy to a certain extent, the circumferential strain acquisition still required the use of bonded-foil gauges for measurement, and measurement errors caused by strain localization still cannot be avoided.

(3) The connection of the strain gauge wires in the Hoek–Franklin triaxial pressure chamber are through the hole reserved at the lower steel platen, rather than some special wire plugs. Accordingly, the wires may have negative impact on oilproof of the rubber sealing sleeve from the rock sample, giving rise to a high rate of test failure.

(4) There are two confining pressure chambers setting on the test machine: the Hoek–Franklin triaxial chamber and self-balancing triaxial pressure chamber. They have to be replaced from each other when conducting different kind of experiments, which greatly increases the labor intensity of laboratory staffs. In addition, the replaced parts are very heavy and high in position, and are prone

to fall during the replacement of the pressure chamber, which may even cause personal injury to the test personnel.

In view of these four major defects mentioned above, after many discussions and consultations with the engineers of the testing machine manufacturer, it was decided to further design and develop the preliminary dynamic disturbance electro-hydraulic servo rock triaxial testing system. The main thought of the secondary modifications to this triaxial test system is to implement the function of applying disturbance load directly in the self-balancing triaxial chamber, which can reduce the labor intensity. At the same time, the axial and circumferential strains can be directly measured by the axial extensometer and the circumferential chain ring extensometer (LVDTs) (as shown in Fig. 2.6a, b) used in conventional triaxial tests. The axial extensometer (LVDT) is directly installed at the position of the upper and lower platens close to the two end faces of the rock sample, which can measure the axial strain of the rock sample more accurately. The circumferential chain ring extensometer (LVDT) is placed at half the height of the rock sample, and the collected strain is the sum of the entire rock sample in the ring direction, which will not cause inaccurate measurement due to the strain localization. As shown in Fig. 2.6c, the axial and circumferential LVDTs are placed outside the rock sample seal layer, the strain gauge connection wires are made of special oil-proof insulated wires, and the wire plugs are directly connected to the sockets reserved on the chassis of the confining pressure chamber. In short, the secondary modification scheme perfectly solves the defects in the preliminary modification scheme, and solves the problems of inaccurate data collection, untight sealing assembly and high labor intensity.

Figure 2.7 shows the photograph and sketch of secondary modification to servo-controlled rock triaxial test system. As can be seen from the figure, the most important modification is that the components related to dynamic disturbance loading and control functions are directly integrated in the main test machine, and the Hoek–Franklin triaxial chamber used for dynamic disturbance test is removed, and the conventional triaxial test and dynamic disturbance triaxial test are completed in the self-balancing triaxial pressure chamber. In terms of details, an external displacement sensor for collecting command and feedback data of dynamic disturbance is installed between the beam of the loading frame and the self-balancing triaxial pressure chamber, and components such as the force sensor (500 kN) for dynamic disturbance and the disturbance rod are directly placed inside the confining chamber. Same as the preliminary modification, the loading rod is also connected to a close loop and controlled by the EDC580 all-digital servo controller independently. After secondary modification, axial disturbance loads can be imposed and controlled independently and precisely during conventional triaxial compression. In addition to that, when conducting disturbance experiments, procedures of rock sample assembly, such as sealing with rubber jacket, installing axial and radial strain extensometers, etc., are exactly same as for conventional triaxial compression experiments. And the axial and circumferential extensometers (LVDTs) are used to measure axial and radial strains during disturbance experiments, which makes the data recorded more accurate than bounded gauges.

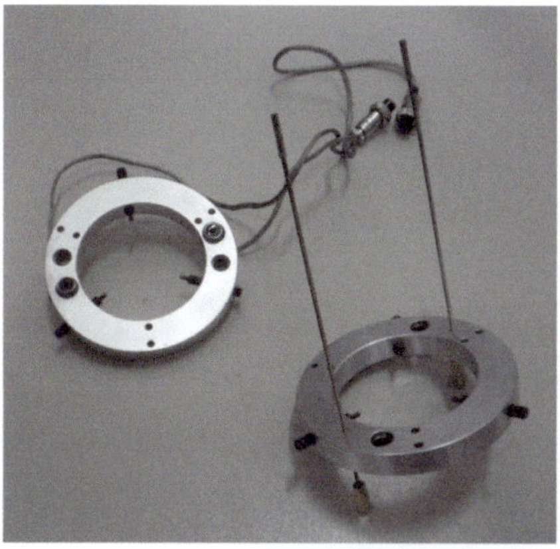

(a) Axial extensometer (LVDT)

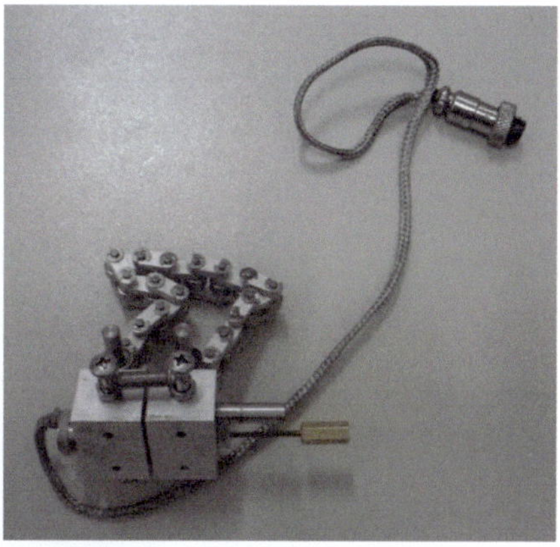

(b) Circumferential extensometer (LVDT)

(c) Installation method of extensometers

Fig. 2.6 Strain gauges and their installation after secondary modification of the triaxial testing system

2.4 Limitation of the Modified Device

It can be seen from the detailed description of the preliminary development and secondary modification of the test equipment mentioned above that the newly modified equipment has capacity to carry out axial dynamic disturbance on rock sample during triaxial compression test. However, there is still a defect in the device that

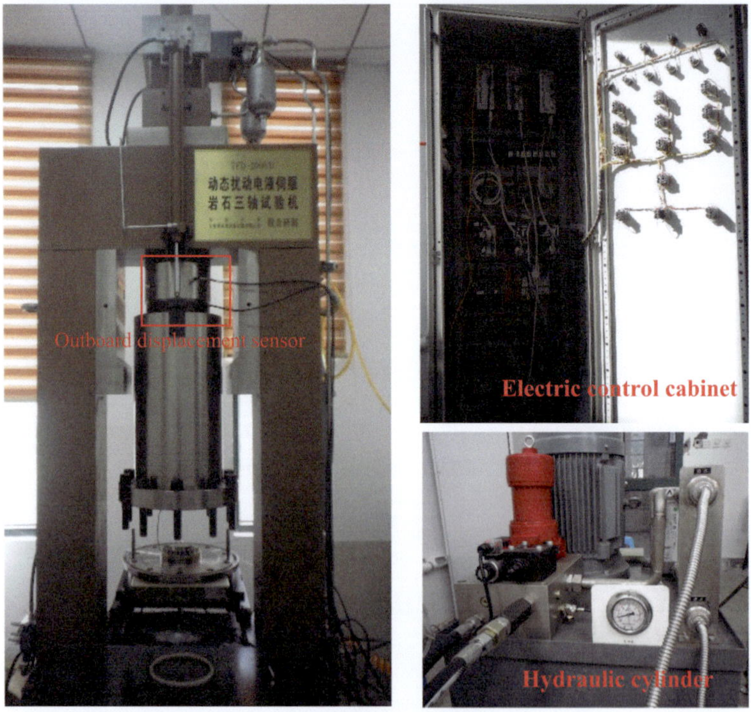

(a) Physical photograph

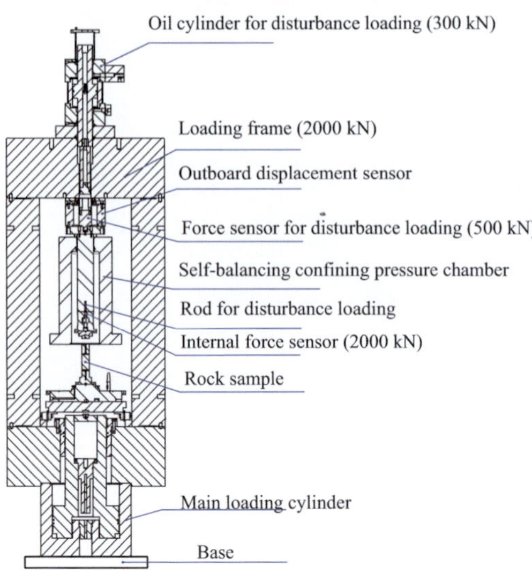

(b) Schematic diagram in front view

Fig. 2.7 Photograph and sketch of secondary modification to servo-controlled rock triaxial test system cannot be fundamentally solved at present, which is explained herein in particular to provide some improvement ideas for later researchers. On the other hand, because of this defect, some scholars may question the reliability of the data obtained by the imperfect testing machine. This special explanation in this section also has the role of clarifying this doubt.

Although the axial loading subsystem and the dynamic disturbance loading subsystem can function independently of each other, these two subsystems will interfere with each other because the loads are applied in the same axial direction of the rock sample. Specifically, in the process of dynamic disturbance test, the axial loading subsystem is usually used to impose load on the end face of the rock sample first, and then the axial stress is stabilized after reaching the target value, that is, the servo control function of the axial loading subsystem is used to lock the axial stress at the target value. After that, the dynamic disturbance loading subsystem is used to carry out the axial dynamic disturbance. When the dynamic disturbance load with a certain amplitude and frequency is applied to the end face of the rock sample, it will cause the fluctuation of the axial servo-control system, that is, the axial loading subsystem will weaken the dynamic disturbance effect to a certain extent, but this weakening is actually only in amplitude, not in frequency.

In order to find out the weakening degree of the axial load to the dynamic disturbance and separate the specific value of dynamic disturbance load in the final measured data, the author has carried out a great load of experimental research work on this issue. For example, we tried to paste some strain gauges on the loading frame to measure its axial deformation, but the measurement results showed that the deformation of the loading frame was extremely small and almost negligible. According to the analysis, it may be due to the small amount of dynamic disturbance load applied under normal circumstances (less than one-tenth of the strength of the rock), and the weakening amount of dynamic disturbance load is even smaller, and at the same time, the stiffness of the force frame is particularly large (much larger than the stiffness of the rock), resulting in a small deformation of the measured loading frame.

After many attempts, it is still impossible to accurately measure the weakening amount, which is the main shortcoming of the developed test equipment. However, after analysis, we still believe that the device can be used in the study of rock damage, deformation and failure caused by dynamic disturbance, for the following reasons:

(1) In the study of dynamic disturbance triaxial test, the main purpose is to explore the mechanical response of rock samples under high stress (axial stress and confining pressure) coupling with slight dynamic disturbance. The applied dynamic disturbance load is usually less than one tenth of the rock's triaxial compressive strength. For example, in the dynamic disturbance uniaxial loading test conducted in Chap. 4 below, the maximum dynamic disturbance load is 5.0 MPa, and the corresponding weakening amount is even smaller, while the uniaxial compressive strength of the red sandstone used is 64.7 MPa.
(2) The sampling frequency of the PID control in axial loading subsystem and dynamic disturbance loading subsystem are quite different, which means that the servo feedback of the axial loading subsystem is ten times slower than that

of the dynamic disturbance loading subsystem. Specifically, in the PID closed-loop control of axial loading subsystem, the sampling frequency is relatively low. Only 3 or less data points per second are collected, which are used to carry out the cyclic process of collection, calculation, feedback and control. However, in the PID closed-loop control process of dynamic disturbance load, the sampling frequency is relatively large. Taking the sine wave dynamic disturbance of 10 Hz as an example, ten cycles of data are required per second. To characterize and feedback the sine wave, five or more data need to be collected per cycle to achieve the purpose, that is to say, dynamic disturbance loading and control requires at least about 50 data per second to complete the cyclic process of acquisition, calculation, feedback and control. Therefore, when the dynamic disturbance test is carried out, the relative attenuation of the amplitude of the dynamic disturbance load will be reduced by the axial loading and the servo slugginess of the control system. That is, the higher the frequency setting in the dynamic disturbance test, the smaller the percentage of the weakening relative to the set amplitude.

To this end, the influence between the dynamic disturbance loading subsystem and the axial loading subsystem of the developed equipment was tested with the applied amplitude of the dynamic disturbance ranging from 1.0 to 20.0 MPa. The collected amplitude data corresponding to the applied amplitude is recorded using the data collection and recording subsystem to calculate the weakening amount and the percentage of the weakening amount relative to the applied amplitude, which are listed in Table 2.1. Figure 2.8 shows the relationship between the percentage of the weakening amount and the applied amplitude of dynamic disturbance. At present, in order to overcome this defect of amplitude reduction, the measured data and fitting curve in Fig. 2.8 have been used as the basic data of the equipment, which is considered in the specific test. The applied amplitude is appropriately increased to make the actual collected amplitude reach the set value of the specific test as far as possible.

2.5 Stability and Reliability Verification of Modified Equipment

2.5.1 Rock Sample Characteristics and Test Verification

For the newly developed testing machine, stability and reliability are the key to the success of the test. In addition to the repeated debugging of the basic design index, the performance of the equipment is also verified by specific dynamic disturbance tests.

As shown in Fig. 2.9, the rock material selected for the verification test is white marble. In accordance with the size recommended by the International Society of Rock Mechanics (ISRM), the rock samples are machined as circular cylinders with

Table 2.1 Test data of interaction effect between dynamic disturbance system and axial loading system

Applied amplitude of dynamic disturbance (MPa)	Collected amplitude of dynamic disturbance (MPa)	Absolute weaking amount (MPa)	Relative weaking amount (%)
1.0	0.544	0.456	45.60
2.0	1.186	0.814	40.70
3.0	1.949	1.051	35.03
4.0	2.752	1.248	31.20
5.0	3.590	1.41	28.20
6.0	4.439	1.561	26.02
7.0	5.222	1.778	25.40
8.0	6.153	1.847	23.09
9.0	7.108	1.892	21.02
10.0	7.961	2.039	20.39
11.0	8.824	2.176	19.78
12.0	9.716	2.284	19.03
13.0	10.617	2.383	18.33
14.0	11.592	2.408	17.20
15.0	12.477	2.523	16.82
16.0	13.371	2.629	16.43
17.0	14.282	2.718	15.99
18.0	15.228	2.772	15.40
19.0	16.158	2.842	14.96
20.0	17.136	2.864	14.32

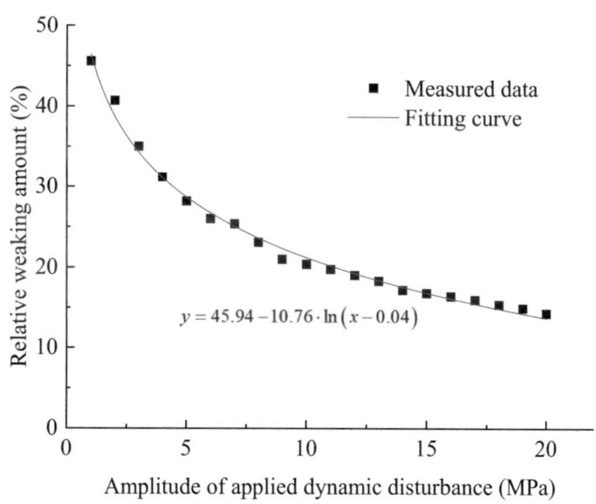

Fig. 2.8 Relationship between applied amplitude and collected amplitude of the modified dynamic disturbance triaxial test system

Fig. 2.9 Marble samples achieved in laboratory

the ratio of height to diameter of 2:1 (height × diameter = 100 mm × 50 mm). The two end faces of the rock samples are polished until the unevenness does not exceed 0.05 mm. Through color observation, density comparison, ultrasonic wave velocity test and porosity analysis, the heterogeneity of the prepared rock samples can be eliminated to ensure the approximate uniformity of each rock sample. The results of multiple uniaxial compression tests show that the average uniaxial compressive strength (UCS) of the selected marble samples is around 49.3 MPa.

During the test, the confining pressure chamber was first filled with silicon oil, and the confining pressure was controlled to the set value by the TEST software. Then the axial load was increased and the data was recorded until the target value was loaded. After that, the dynamic disturbance load was controlled to impose on the rock sample by the software DOLI Installation Center. According to the requirements of the test, at a certain level of axial stress and radial stress, a certain amplitude and frequency of disturbance load can be applied to load the sample to failure, and the influence of dynamic disturbance on the mechanical properties of the rock during the loading process can be studied. The axial load or confining pressure can also be gradually removed when the disturbance load is applied, and the mechanical properties of the rock under the synergistic action of dynamic disturbance and unloading can be observed. Of course, when dynamic disturbance is performed, disturbance loads of different frequencies and amplitudes can also be applied in different time periods to study the mechanical properties of rocks under various kinds of disturbances.

This test mainly considers the influence of the amplitude of the dynamic disturbance load on the mechanical properties of the marble. The rock sample was first uniaxial compressed to 50 kN (about 50% of the uniaxial compressive strength, 25 MPa), and then the dynamic disturbance loads of the same frequency (5 Hz) but with different amplitudes were applied until the rock samples failed.

2.5.2 Verification Result

Figure 2.10 shows the variation curves of axial loads acting on rock samples with time with disturbance amplitudes of 1.2, 5.5, and 12 kN. As can be seen from the figure, when the frequency of dynamic disturbance is constant, the time from beginning of disturbance to failure of rock sample is inversely correlated with the disturbance amplitude. A greater amplitude makes the rock samples experience a shorter disturbance time, and vice versa. In other words, when the rock sample is under a certain level of axial stress, the external cyclic load is applied to it and eventually leads to failure of rock sample. If the frequency of the disturbance is fixed, the frequency of the cyclic disturbance load is inversely correlated with its amplitude. Therefore, it can be concluded that the action process of dynamic disturbance is actually a cumulative process of constant superposition of damage. In this process, the axial deformation of the rock sample increases continuously, and the axial deformation is mainly caused by three parts: the axial stress in the target level, the dynamic disturbance, and the creep caused by the first two with the increase of time. From the perspective of stress level alone, the maximum stress after dynamic disturbance does not reach the stress during uniaxial compression failure. Therefore, the failure criterion derived from the stress cannot effectively determine whether the rock sample will be damaged under the action of dynamic disturbance. What kind of failure criterion should be selected for rock damage and failure under dynamic disturbance, and whether new failure criterion should be established are all important problems to be solved in future research.

Figure 2.11 shows the failure modes of rock samples under four different test conditions. It can be seen from this figure that the marble sample shows splitting failure under uniaxial compression without dynamic disturbance, and the cracks mainly follow the direction of the principal stress because of Poisson effect. It is found by comparison that the rock samples induced by dynamic disturbance with frequency of 5 Hz and three different amplitudes show a fragmented failure pattern, and the degree of fragmentation increases with the increase of the amplitude of disturbance load. Accordingly, from the load-time curves and failure modes obtained from the test, the developed dynamic disturbance rock servo triaxial test system has high stability and reliability.

2.6 Chapter Summary

This chapter describes in detail the preliminary development and secondary modification of dynamic disturbance rock servo triaxial test system. The modified equipment can independently control and apply axial stress, confining pressure, as well as dynamic disturbance. The equipment developed for the first time is mainly to add a disturbance rod in the upper platen of the Hoek–Franklin triaxial pressure chamber.

Fig. 2.10 Load-time curves in the condition of dynamic disturbance load on same frequency but different amplitudes

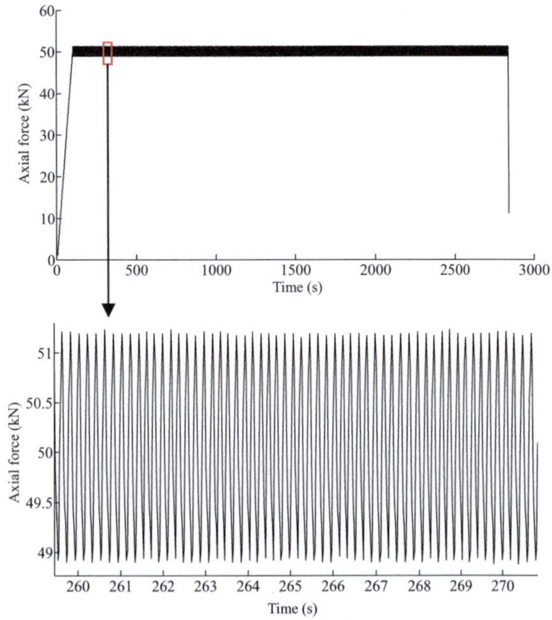

(a) Frequency of 5 Hz, amplitude of 1.2 kN

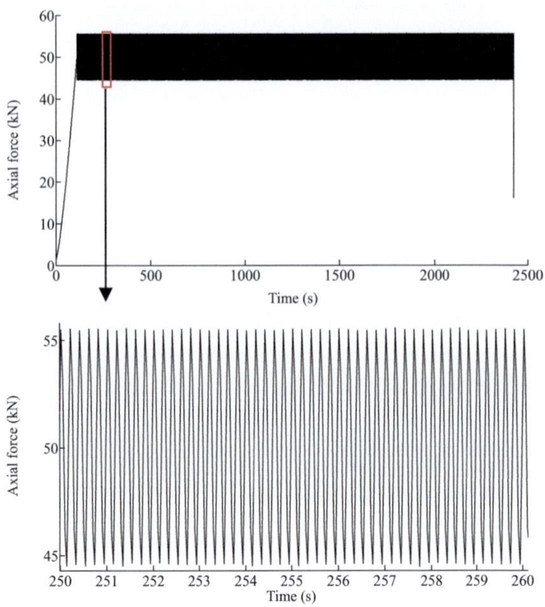

(b) Frequency of 5 Hz, amplitude of 5.5 kN

Fig. 2.10 (continued)

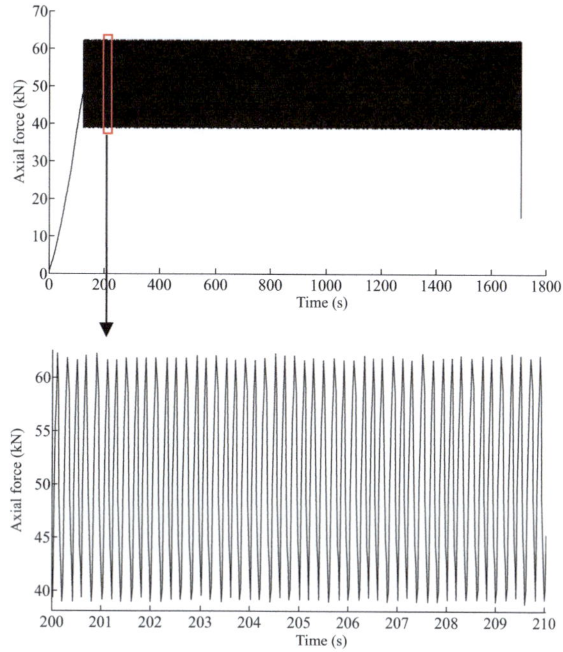

(c) Frequency of 5 Hz, amplitude of 12 kN

The disturbance rod is connected with the disturbance cylinder, load sensor, displacement sensor, EDC580 servo controller, and computer, enabling dynamic disturbance with certain frequencies and amplitudes to be applied simultaneously with axial stress and confining pressure. The developed equipment can be used to study the mechanical response of rock damage, deformation and failure when they are subjected to high stress levels (using confining pressure and axial loading subsystems) and external disturbance loads (using dynamic disturbance loading subsystem).

Due to some defects and limitations of the preliminarily developed equipment, secondary modification to the dynamic disturbance rock servo triaxial test system is conducted. The main thought of the secondary modifications to this triaxial test system is to remove Hoek–Franklin triaxial pressure chamber and implement the function of applying disturbance load directly in the self-balancing triaxial chamber, which can reduce the labor intensity. At the same time, the axial and circumferential strains can be directly measured by the axial extensometer and the circumferential chain ring extensometer (LVDTs), solving the problems such as unreliable sealing, measurement distortion due to strain localization, etc., and making the data recorded more accurate.

This chapter also verifies the stability and reliability of the developed equipment through a set of tests, the purpose of which is to observe the mechanical response of the marble samples under the coupling action of uniaxial loading and dynamic disturbance. The test results show that when the frequency of dynamic disturbance

Fig. 2.11 Failure modes of rock samples

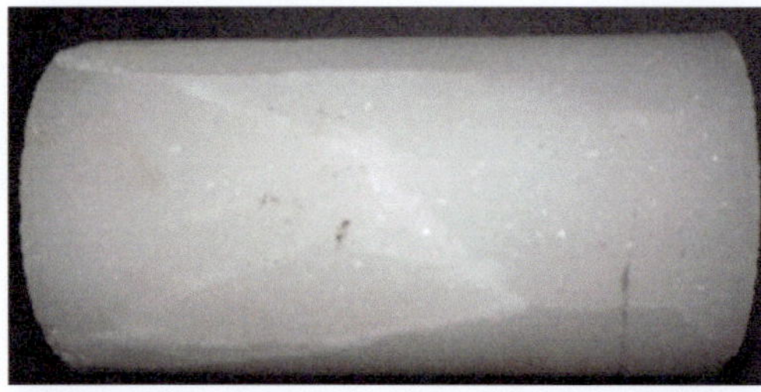

(a) Uniaxial compression with no dynamic disturbance

(b) Frequency of 5 Hz, amplitude of 1.2 kN

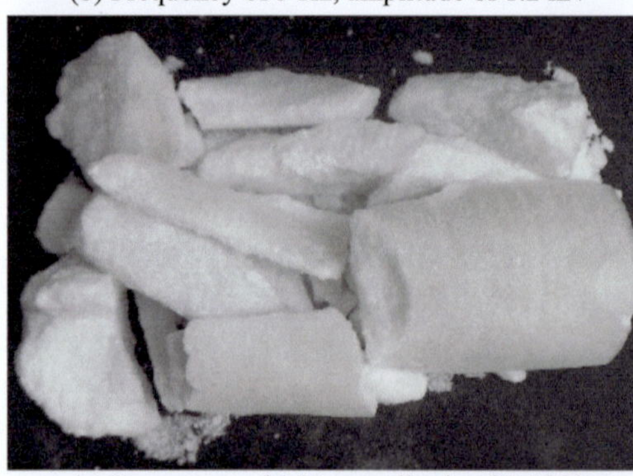

(c) Frequency of 5 Hz, amplitude of 5.5 kN

(d) Frequency of 5 Hz, amplitude of 12 kN

is constant, the time from beginning of disturbance to failure of rock sample is inversely correlated with the disturbance amplitude. A greater amplitude makes the rock samples experience a shorter disturbance time, and vice versa. The failure mode under dynamic disturbance is very different from that under conventional uniaxial compression. The failure mode after dynamic disturbance is more fragmented, and the degree of fragmentation increases with the increase of disturbance amplitude. It is indicated that the developed dynamic disturbance rock servo triaxial test system has high stability and reliability.

References

Azra NT, Augusto LP, Mukul MS (1998) Nonlinear viscoelastic behavior of sedimentary rocks; Part II, Hysteresis effects and influence of type of fluid on elastic moduli. Geophysics 63(1):195–203. https://doi.org/10.1190/1.1444313

Erarslan N, Williams DJ (2012) The damage mechanism of rock fatigue and its relationship to the fracture toughness of rocks. Int J Rock Mech Min Sci 56:15–26. https://doi.org/10.1016/j.ijrmms.2012.07.015

Franklin JA, Hoeck E (1970) Developments in triaxial testing technique. Rock Mech 2:223–228. https://doi.org/10.1007/BF01245576

Fuenkajorn K, Phueakphum D (2010) Effects of cyclic loading on mechanical properties of Maha Sarakham salt. Eng Geol 112:43–52. https://doi.org/10.1016/j.enggeo.2010.01.002

Gong FQ, Lu DH, Li XB et al (2014) Toughness increasing or decreasing effect of hard rock fracture with pre-static loading under dynamic disturbance. Chin J Rock Mech Eng 33(9):1905–1915 (in Chinese). https://doi.org/10.13722/j.cnki.jrme.2014.09.022

Goodman RE (1989) Introduction to rock mechanics, 2nd edn. Wiley, Hoboken

He MC, Cheng T, Qiao YF et al (2023a) A review of rockburst: experiments, theories, and simulations. J Rock Mech Geotech Eng 15(5):1312–1353. https://doi.org/10.1016/j.jrmge.2022.07.014

He BG, Wang L, Feng XT et al (2023b) Failure modes of jointed granite subjected to weak dynamic disturbance under true-triaxial compression. Rock Mech Rock Eng 56:7939–7957. https://doi.org/10.1007/s00603-023-03507-9

Hoek E (1968) Simple triaxial cell for field or laboratory testing of rock. Trans Inst Min Metall 77:A22–A26

Jaeger JC (1967) Brittle fracture of rocks in failure and breakage of rock. In: Fairhurst C (ed). AIME, New York

Li XB, Li DY, Guo L et al (2007) Study on mechanical response of highly-stressed pillars in deep mining under dynamic disturbance. Chin J Rock Mech Eng 26(5):922–929 (in Chinese)

Liu JF, Xu J, Li QS et al (2010) Experimental research on damping parameters of rock under cyclic loading. Chin J Rock Mech Eng 29(5):1036–1041 (in Chinese)

Liu F, Ma TH, Tang CA et al (2018) Prediction of rockburst in tunnels at the Jinping II hydropower station using microseismic monitoring technique. Tunn Undergr Space Technol 81:480–493. https://doi.org/10.1016/j.tust.2018.08.010

Ma LJ, Liu XY, Wang MY et al (2013) Experimental investigation of the mechanical properties of rock salt under triaxial cyclic loading. Int J Rock Mech Min Sci 62(9):34–41. https://doi.org/10.1016/j.ijrmms.2013.04.003

Naji AM, Emad MZ, Rehman H et al (2018) Geological and geomechanical heterogeneity in deep hydropower tunnels: a rock burst failure case study. Tunn Undergr Space Technol 84:507–521. https://doi.org/10.1016/j.tust.2018.11.009

Paterson MS, Wong TF (2005) Experimental rock deformation—the brittle field. Springer, Berlin. https://doi.org/10.1007/b137431

Song DZ, Wang EY, Liu J (2012) Relationship between EMR and dissipated energy of coal rock mass during cyclic loading process. Saf Sci 50:751–760. https://doi.org/10.1016/j.ssci.2011.08.039

Yang S, Ning JG, Gao MT et al (2023) Instability mechanism and stability control of gob-side entry in a deep mine: a case study. Bull Eng Geol Environ 82(346). https://doi.org/10.1007/s10064-023-03362-6

Zhou ZL, Li GN, Ning SL et al (2014) Acoustic emission characteristics and failure mechanism of high-stressed rocks under lateral disturbance. Chin J Rock Mech Eng 33(8):1720–1728 (in Chinese). https://doi.org/10.13722/j.cnki.jrme.2014.08.023

Zhu ZD, Sun LZ, Wang MY (2010) Damping ratio experiment and mesomechanical analysis of deformation failure mechanism on rock under different frequency cyclic loadings. Rock Soil Mech 31(S1):8–12 (in Chinese). https://doi.org/10.16285/j.rsm.2010.s1.062

Zuo YJ, Li XB, Tang CA et al (2006) Experimental investigation on failure of rock subjected to 2d dynamic-static coupling loading. Chin J Rock Mech Eng 25(9):1809–1809 (in Chinese)

Chapter 3
Description of Rock Samples and Their Basic Physical and Mechanical Properties

3.1 Introduction

Rock materials are natural materials given to us by nature, which are formed by different diagenesis and exist on the earth's surface under complex geological conditions over a very long geological time. The different diagenesis and complex occurrence environment also lead to different types of rock materials have a great difference in physical and mechanical properties, showing a very significant non-uniformity, non-continuity, nonlinear characteristics. In view of this, before the mechanics tests, it is necessary to select a type or several types of rocks suitable for the study of specific problems in advance, and to determine their physical and mechanical properties with the purpose of initially understanding the basic properties of the selected rock materials. It can not only help us understand the basic situation of the selected rock materials, but also provide some basic reference data for the subsequent complex mechanical tests.

From the perspective of meso-level, the wide and random distribution of pores and micro-cracks in the rock, as well as different mineral particles and cementation forms, lead to the high non-uniformity of the rock. However, the cyclic dynamic disturbance tests in this study do not take the non-uniformity of rock as the influencing factor for discussion. On the contrary, the rock samples used in this study are required to be as consistent and uniform as possible at the macro level. Therefore, sandstone, which is relatively common, easy to process and relatively uniform, is selected as the test material, and macroscopic cracks in the rock samples are also avoided as much as possible in the preparation such as core drilling, cutting, end grinding. The determination of basic physical properties can further eliminate the rock samples with uneven texture. The measurement of density, porosity, P- and S-wave velocity, and other physical quantities can also comprehensively and quantitatively assess the consistency of rocks from different angles, so as to screen rock samples for subsequent dynamic disturbance tests.

X. Chen, *Damage and Failure of Rock Triggered by Dynamic Disturbance*,
https://doi.org/10.1007/978-981-96-9486-0_3

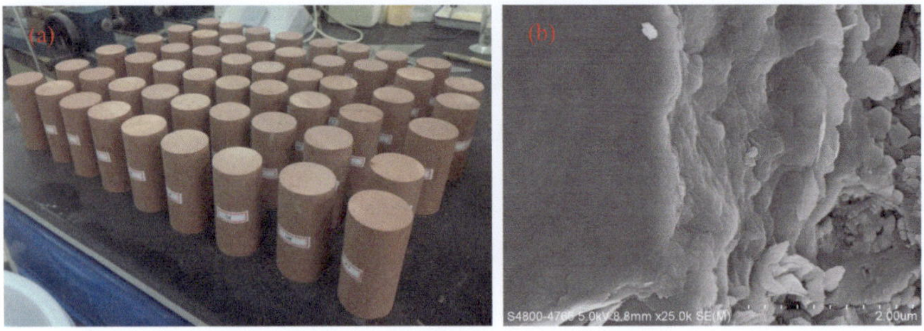

Fig. 3.1 Sandstone samples used in this research: **a** picture of the prepared sandstone samples, **b** microscopic structure observed using a scanning electron microscope (SEM)

3.2 Rock Sample Preparation

Red sandstone samples, which were procured from a quarry in Zigong City, Sichuan Province, China, were used in the experiments. This type of rock was chosen because of its high degree of sample consistency and homogeneity. The samples were a fine-grained heterogeneous rock with a mean grain size of around 0.14 mm. From X-ray diffraction (XRD) analysis, the mineralogical composition of the sandstone is approximately dominated by feldspar (~ 46%), quartz (~ 21%), kaolinite (~ 14%), chlorite (~ 9%), illite (~ 7%) and others (~ 3%).

In accordance with the International Society of Rock Mechanics (ISRM) suggested standards for intact rock testing (Franklin 1983), the rock samples were cored, cut and polished into cylindrical shapes with height of 100 mm and diameter of 50 mm. The height-to-diameter ratio of the samples were 2:1. To minimize the difference between rock samples caused by origination, all the cores were drilled from a single intact big rock block without any macroscopic cracks. Photograph of the prepared samples and a representative microscopic structure are displayed in Fig. 3.1.

3.3 Basic Physical and Mechanical Properties of the Rock Samples

There are two main purposes for the determination of the basic physical parameters of the rock samples: (1) to recognize and understand the basic characteristics of the selected sandstone, such as the apparent density, porosity, S- and P-wave velocity; (2) to screen out the rock samples with relatively uniform properties and small differences among them and to minimize the errors caused by the differences for subsequent cyclic dynamic disturbance triaxial tests.

3.3.1 Apparent Density

The size of the rock sample (diameter d and height H) is mainly measured using a vernier caliper, and the apparent volume V of the rock samples can be calculated by using the measured rock sample size

$$V = \frac{1}{4}\pi d^2 \cdot H. \tag{3.1}$$

Before the density determination is adopted, the rock sample must be dried at 100 °C for 24 h in the dryer. The mass m of the rock sample is weighed with a precision electronic balance after the rock sample is cooled naturally. The rock sample density ρ is finally calculated by using the formula:

$$\rho = \frac{m}{V}. \tag{3.2}$$

The apparent density test results for the individual rock samples used for the tests are listed in Table 3.1.

3.3.2 Porosity

As shown in Fig. 3.2a, the KXD-II rock core porosimeter manufactured by Huaxing Petroleum Instrument Co., Ltd. is employed to determine porosity of each rock sample. Using the principle of gas expansion, the skeleton volume of the rock cores is measured. Specifically, turn on the valve in Fig. 3.2b to make the helium in the known chamber with pressure of P_k and volume of V_k expand to the adjacent unknown chamber under constant temperature, and then observe the scale on the pressure meter to determine the equilibrium pressure (P). Using Boyle's gas law

$$P_k V_k = PV + PV_k, \tag{3.3}$$

the unknown chamber volume (V) can be calculated by

$$V = \frac{P_k V_k - PV_k}{P}. \tag{3.4}$$

Since the wall has a certain pressure property, Eq. (3.4) can be modified as

$$V = \frac{P_k V_k - PV_k}{P} + \frac{P_k - P}{P} \cdot G \cdot (P_0 + P), \tag{3.5}$$

where G represents the pressure variation coefficient of the wall, $cm^3\ MPa^{-1}$; P_0 is atmospheric pressure, MPa.

Table 3.1 Basic physical parameters of rock samples

Sample no.	Size	Bulk density (g cm^{-3})	Connected porosity (%)	P-wave velocity (m s^{-1})	S-wave velocity (m s^{-1})
	Height × diameter (mm × mm)				
RS-1-1	99.64 × 49.88	2.361	5.82	2985.3	1947.2
RS-1-2	100.00 × 49.82	2.335	5.40	3031.9	1994.2
RS-1-3	99.34 × 49.90	2.337	5.72	2908.8	2032.7
RS-2-1	97.86 × 49.84	2.354	5.18	2939.5	1938.0
RS-2-2	99.00 × 49.86	2.343	5.66	2969.1	2011.3
RS-2-3	98.60 × 49.92	2.352	5.81	3000.9	1970.5
RS-3-1	99.56 × 49.92	2.331	4.91	2883.4	1841.2
RS-3-2	98.68 × 49.84	2.349	5.29	2910.0	1969.0
RS-3-3	100.08 × 49.88	2.351	5.38	3066.3	1994.7
RS-4-1	98.72 × 49.86	2.362	5.63	2925.5	2043.3
RS-4-2	99.82 × 49.94	2.364	5.50	3063.1	1882.9
RS-4-3	99.06 × 49.84	2.375	5.97	3032.5	2061.1
RS-5-1	99.66 × 49.92	2.355	6.02	2897.2	2015.3
RS-5-2	100.40 × 49.86	3.341	5.04	3031.0	2031.7
RS-5-3	99.50 × 49.90	2.331	5.73	2969.0	1982.1
RS-6-1	98.64 × 49.88	2.356	4.90	2015.6	2161.5
RS-6-2	99.04 × 49.94	2.343	5.77	3093.2	2026.4
RS-6-3	98.76 × 49.86	2.339	5.49	3094.2	1983.4
RS-7-1	98.62 × 49.90	2.362	5.65	3110.5	2058.6
RS-7-2	99.10 × 49.82	2.364	5.61	2910.1	2048.3
RS-7-3	99.34 × 49.96	2.351	5.85	2938.9	1991.8
RS-8-1	99.36 × 49.94	2.367	5.74	3178.0	1910.4
RS-8-2	99.64 × 49.84	2.344	5.34	3094.2	1993.2
RS-8-3	99.28 × 49.88	2.339	5.58	3226.3	2048.7
RS-9-1	99.40 × 49.88	2.358	5.49	2868.4	1949.2
RS-9-2	98.74 × 49.90	2.308	5.09	2911.0	2031.6
RS-9-3	98.66 × 49.86	2.349	5.81	2880.9	1952.3
RS-10-1	100.00 × 49.90	2.354	5.26	3015.5	1988.4
RS-10-2	99.34 × 49.92	2.340	5.44	3077.2	1921.3
RS-10-3	99.00 × 49.86	2.366	5.57	3181.0	1951.4
RS-11-1	98.60 × 49.96	2.341	5.61	2970.1	1939.0
RS-11-2	98.68 × 49.90	2.375	5.39	2845.9	1944.7
RS-11-3	100.08 × 49.86	2.371	5.71	3046.5	1979.0
RS-12-1	99.60 × 49.88	2.358	5.65	3032.0	1936.5

(continued)

Table 3.1 (continued)

Sample no.	Size Height × diameter (mm × mm)	Bulk density (g cm^{-3})	Connected porosity (%)	P-wave velocity (m s^{-1})	S-wave velocity (m s^{-1})
RS-12-2	99.34 × 49.92	2.362	5.48	3110.6	1955.6
RS-12-3	99.06 × 49.90	2.375	5.62	3015.4	1938.7
RS-13-1	99.28 × 49.94	2.364	5.80	2910.3	1947.2
RS-13-1	100.04 × 49.86	2.367	5.59	3049.7	1983.1
RS-13-3	99.50 × 49.88	2.338	5.61	2868.4	1959.2
Average	–	2.3777	5.541	2975.83	1982.43

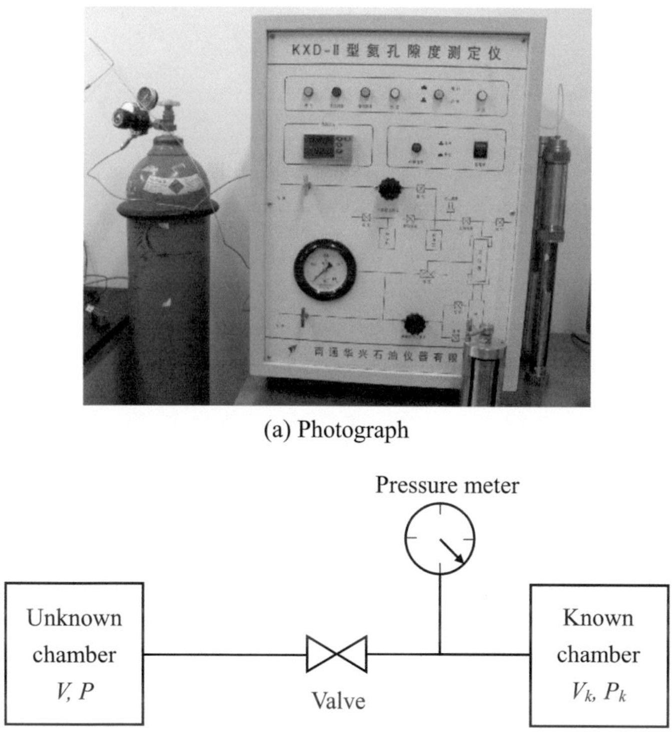

(a) Photograph

(b) Schematic diagram

Fig. 3.2 KXD-II core porosimeter

In order to determine the volume of the rock skeleton, the above procedure needs to be repeated twice. In the first test, the volume of the unknown chamber with no rock sample loaded (V_1) can be measured

$$V_1 = \frac{P_k V_k - P_1 V_k}{P_1} + \frac{P_k - P_1}{P_1} \cdot G \cdot (P_0 + P_1), \tag{3.6}$$

where P_1 represents the equilibrium pressure measured when the rock sample is not loaded.

In the second test, the rock sample is loaded into the unknown chamber, and the volume of the unknown chamber with a rock sample loaded (V_2) can be measured in a same way

$$V_2 = \frac{P_k V_k - P_2 V_k}{P_2} + \frac{P_k - P_2}{P_2} \cdot G \cdot (P_0 + P_2), \tag{3.7}$$

where P_2 represents the equilibrium pressure measured when a rock sample is loaded. It is note that V_2 includes the volume of connected pores within the rock sample.

Therefore, the volume difference obtained from the two measurements can be used to calculate the skeleton volume of the rock V_S

$$V_S = V_1 - V_2. \tag{3.8}$$

Finally, the connected porosity of the rock sample (ϕ) can be determined using the following formula

$$\phi = \frac{V - V_S}{V}. \tag{3.9}$$

The porosity test results of each rock sample in this test are summarized in Table 3.1.

3.3.3 *Ultrasonic Wave Velocity*

As illustrated in Fig. 3.3, the ultrasonic wave velocities were measured with an assembly device which consists of a Tektronix digital phosphor oscilloscope 3012, an Olympus 5077PR square wave pulser-exciter and two P-wave and two S-wave transducers with a frequency of 0.5 MHz.

During the test, an appropriate amount of petroleum jelly is first applied to both ends of the rock sample, in order to make the acoustic transducer and the rock sample better coupled and reduce the time delay caused by the gap between the transducers and the rock sample. Then, the acoustic wave is excited by the acoustic wave exciter (using a single excitation), and the time difference between the excited acoustic wave

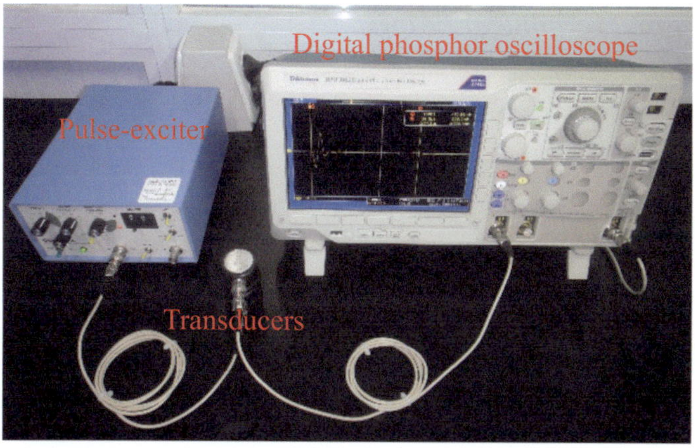

Fig. 3.3 Assembled setup for testing acoustic wave velocity

and the first wave starting point of the oscilloscope is checked on the digital phosphor oscilloscope, and the shear wave velocity and longitudinal wave velocity of the rock sample (v) are calculated by the ratio of the sample height to the time difference

$$v = \frac{H}{t_1 - t_0}, \tag{3.10}$$

where H is the distance of the sound wave propagation, that is, the height of the rock sample; t_0 is the time point of the sound wave excitation, t_1 is the time point of the first wave reading on the oscilloscope; and $t_1 - t_0$ represents the time taken by the sound wave to propagate in the rock sample.

The test results of shear wave velocity and longitudinal wave velocity of each rock sample are also summarized in Table 3.1.

In summary, by measuring the five parameters of each rock sample, such as size, apparent density, porosity, shear wave velocity and longitudinal wave velocity, rock samples with large differences from the mean value (relative error greater than 10%) can be effectively eliminated, and rock samples with relatively similar physical properties can be selected for subsequent triaxial dynamic disturbance tests. The screened rock samples are numbered in the type of RS-X-Y, in which RS is the abbreviation of red sandstone, and the middle number X is the numbering sequence. Since each test condition needs to be repeated at least three times, the end number Y is used to indicate the serial number of repeated tests under the same test condition. The average apparent density and porosity of the numbered samples are 2.3777 g cm^{-3} and 5.541%. The average velocity of longitudinal wave and shear wave are 2975.83 m s^{-1} and 1982.43 m s^{-1}, respectively.

3.3.4 Basic Mechanical Properties

In addition to the determination of the basic physical properties of the rock samples used, conventional triaxial compression tests under five different levels of confining pressure conditions, 0, 10, 20, 30 and 40 MPa, were also carried out according to the method recommended by the International Society of Rock Mechanics (ISRM) (Franklin 1983), aiming to understand the basic mechanical properties of the selected rock (such as triaxial compressive strength, elastic modulus, Poisson's ratio, internal friction angle, cohesion, etc.) and to obtain certain data basis for the subsequent triaxial cyclic dynamic disturbance tests. The equipment used was the conventional electro-hydraulic servo rock triaxial testing system (TFD-2000/D, as described in detail in Sect. 2.2). The axial stress was incrementally increased at a loading rate of 0.02 mm min^{-1} with axial displacement controlling mode, and the development of the deviator stress with the axial and radial strains was collected during the loading process.

Figure 3.4 plots the deviatoric stress–strain curves obtained from the triaxial compressive tests with the confining pressures ranging from 0 MPa (uniaxial compression) to 40 MPa. It can be clearly seen from the figure that the confining pressure effect of the rock, that is, the compressive strength of the rock increases with the increase of confining pressure, and the axial strain and radial strain corresponding to the stress peak point (the negative value only indicates the direction of the strain, and the direction of deformation reduction is positive according to the usual habit of rock mechanics) also increases with the increase of confining pressure, and it shows the phenomenon that the ductility increases with the increase of confining pressure. According to the elastic theory, assuming that the rock is an elastic material, the formula for the relationship between stress, strain, elastic modulus, and Poisson's ratio can be expressed as (Timoshenko and Goodier 1951):

$$E = \frac{(\sigma_1 + \sigma_3)\sigma_1 - 2\sigma_3^2}{(\sigma_1 + \sigma_3)\varepsilon_1 - 2\sigma_3\varepsilon_3}, \tag{3.11}$$

$$\nu = \frac{\sigma_1\varepsilon_3 - \sigma_3\varepsilon_1}{2\sigma_3\varepsilon_3 - (\sigma_1 + \sigma_3)\varepsilon_1}, \tag{3.12}$$

where σ_1 and ε_1 are the axial stress and strain which are taking as values corresponding to half of the maximum deviatoric stress; σ_3 is the confining pressure; ε_3 is the radial strain corresponding to half of the maximum deviatoric stress. The elastic modulus and Poisson's ratio calculated by Eqs. (3.11) and (3.12) at the five different levels of confining pressures are listed in Table 3.2, and the Young's modulus E and Poisson's ratio ν of the sandstone selected for the test are determined by their average values to be 12.735 GPa and 0.310, respectively.

As shown in Fig. 3.5, the triaxial compressive strengths under the five different levels of confining pressures can also be used to plot the stress Mohr circles and strength envelope. The first data in the legend brackets denotes the confining pressure, and the second data is the triaxial compressive strength of the rock measured under

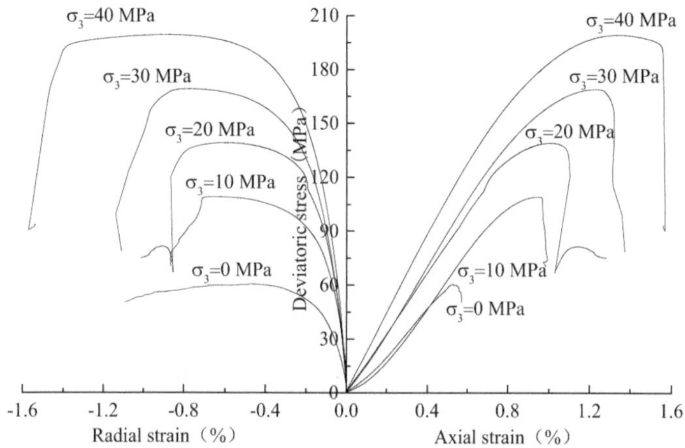

Fig. 3.4 Deviatoric stress versus axial and radial strain obtained from the monotonous triaxial compression tests

Table 3.2 Calculated Young's modulus, Poisson ratios and their average values under five different levels of confining pressures

Confining pressure (MPa)	Triaxial compressive strength (MPa)	Elastic modulus (GPa)	Poisson's ratio
0	64.70	10.826	0.214
10	109.17	11.020	0.253
20	139.12	13.540	0.329
30	168.99	13.178	0.366
40	199.25	15.111	0.388
Average	–	12.735	0.310

the corresponding confining pressure, which is taken as the peak deviatoric stresses $(\sigma_1 - \sigma_3)$. The cohesion c and the internal friction angle φ are determined as 16.2 MPa and 39° by employing the linear Mohr–Coulomb criterion as follows:

$$\sigma_1 - \sigma_3 = \frac{2c\cos\phi}{1 - \sin\phi} + \frac{2\sin\phi}{1 - \sin\phi}\sigma_3. \tag{3.13}$$

Figure 3.6 shows the failure modes of red sandstone samples under five different levels of confining pressures. It can be seen from the figure that the rock samples all show single shear band and brittle failure characteristics under uniaxial and triaxial compressive conditions.

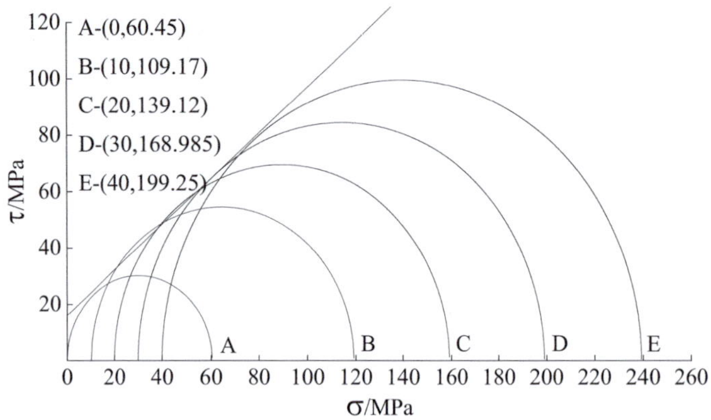

Fig. 3.5 Mohr stress circles

Fig. 3.6 Failure modes of rock samples under conventional triaxial compression

3.4 Chapter Summary

This chapter mainly introduces the preparation of rock samples and the testing of basic physical and mechanical properties of each rock sample before the triaxial cyclic dynamic disturbance tests.

The basic physical properties of the rock sample were tested in terms of apparent density, porosity, S- and P-wave velocity. The results show that the average apparent density and porosity of the numbered samples are 2.3777 g cm^{-3} and 5.541%. The average velocity of longitudinal wave and shear wave are 2975.83 m s^{-1} and 1982.43 m s^{-1}, respectively.

The basic mechanical properties of rock samples are tested using a rock triaxial testing system. The triaxial compression strengths of the sandstone are respectively

60.45 MPa, 109.17 MPa, 139.12 MPa, 168.99 MPa, and 199.25 MPa under the confining pressures of 0 MPa, 10 MPa, 20 MPa, 30 MPa, and 40 MPa. The Young's modulus E and Poisson's ratio v of the sandstone are respectively calculated as approximately 12.74 GPa and 0.31, respectively. The cohesion c and the internal friction angle φ are determined as 16.2 MPa and 39° by employing the linear Mohr–Coulomb criterion.

References

Franklin JA (1983) Suggested methods for determining the strength of rock materials in triaxial compression: revised version. Int J Rock Mech Min Sci Geomech Abs 20(6):285–290. https://doi.org/10.1016/0148-9062(83)90598-3

Timoshenko S, Goodier JN (1951) Theory of elasticity. McGraw-Hill Book Company, New York

Chapter 4
Experimental Study on the Influence of Dynamic Disturbance on Progressive Damage and Failure of Rock

4.1 Introduction

With the increase in mineral resources demand and exhaustion of shallow resources, exploitations of deep underground resources have gradually become necessity and deeper excavations have been carried out in recent years (Mazaira and Konicek 2015; Li et al. 2017). However, the loading states of rock in deep-buried excavations have different characteristics from "conventional loading" in the laboratory (Zhou et al. 2014; Gischig et al. 2016). Specifically, the deep-buried rock is generally not in a monotonous loading condition, but may experience external disturbance loads (such as blasting and drilling vibration, overburden fracture, groundwater pressure changes, rock bursts, etc.) and bear overstress simultaneously in the process of free surfaces formation of excavations. Due to the lack of understanding of mechanical behaviors and responses of deep-buried rock under these complex stress conditions, it is difficult to effectively predict, control and prevent various disturbance-triggered engineering accidents and even disasters that may cause fatal injuries to workers and significant loss of equipment and time. Therefore, it is critically necessary to investigate the mechanical behaviors and responses of rocks under high stress in conjunction with external disturbance load for the purpose of safety and stability of deep-buried mining and other deep rock engineering.

The determination of disturbance of rocks in the laboratory has been gaining importance in recent years. Extensive studies were carried out in the past to examine the effect of the cyclic loads on rock strength and deformation characteristics and other mechanical responses. For example, Royer-Carfagni and Salvatore (2000) concentrated their efforts in reproducing the degradation process of three different qualities of marbles by performing low-cycle number, uniaxial compression tests. It was concluded that, even when cyclic loading tests did not reproduce the exact conditions that stone underwent under operating conditions, they might nonetheless constitute an effective way of measuring an intrinsic property of the material. Bagde and Petroš (2005a, b) performed laboratory investigations on intact sandstone samples

X. Chen, *Damage and Failure of Rock Triggered by Dynamic Disturbance*,
https://doi.org/10.1007/978-981-96-9486-0_4

to explore the mechanical properties when exposed to dynamic uniaxial cyclical loads with various loading frequencies, amplitudes and waveforms and found that the loading frequency, amplitude as well as the waveform, was of great significance and influenced the rock behavior in dynamic cyclic loading conditions. Liu et al. (2011, 2012), Liu and He (2012) carried out a series of laboratory tests to study the effect of confining pressure and loading frequency on the mechanical properties and damage evolution of sandstone samples under cyclic loading conditions. Result from the tests indicated that the level of confining pressure had a significant influence on cyclic dynamic deformation and damage evolution of the sandstone samples. Ren et al. (2013), Ma et al. (2013) and Wang et al. (2013) carried out similar cyclic loading experiments on rock salt and granite samples, and they were also of the opinion that fatigue behavior of granite varies with the tendency for volumetric change in triaxial cyclic compression tests. Taheri et al. (2016) performed an experimental investigation on Hawkesbury sandstone samples to study the peak strength variations during triaxial monotonic and cyclic compressive testing. Their results confirmed that beginning of cyclic loading at higher stress will result in failure after fewer cycles, and the mechanical properties of the rock were altered by systematic cyclic loading; dependent upon the applied stress level during cyclic loading the rock may become weaker or stronger as a result of cyclic loading. Fan et al. (2016) conducted comparisons between conventional cyclic loading tests and interval cyclic loading tests, which combine spaced stress cycles and normal stress cycles. Experimental measurements demonstrated that the combined cyclic stress has a strong impact on the fatigue activity of rock salt. In their recent publications Liu et al. (2017, 2018) systematically investigated the mechanical properties of synthetic jointed rock models under different dynamic cyclic conditions with the methods of numerical simulation and laboratory tests. Their experimental results revealed the influence of the three cyclic loading parameters on the mechanical properties of jointed rock models, regarding the fatigue deformation characteristics, the fatigue energy and damage evolution, and the fatigue failure and progressive failure behavior. On the other hand, some numerical methods for simulating damage or fracture of quasi-brittle material subjected to complex loading conditions, such as RFPA (rock failure process analysis) (Tang 1997), XFEM (extended finite element method) (Wang et al. 2016), SBFEM (scaled boundary finite element method) (Bek et al. 2018), strong discontinuity embedded approach and cracking elements methods (Zhang et al. 2015), cracking particles (Rabczuk and Belytschko 2004; Rabczuk et al. 2010), phase field methods and peridynamics (Rabczuk and Ren 2017), which are potential approaches to the progressive damage and failure of rock materials caused by disturbance loads. The above studies chiefly emphasize the deformation characteristics, damage evolution and dynamic properties of rock subjected to cyclic loading, and they have provided acceptable estimations of the response of cyclic disturbance loading with relatively lower frequencies for rock samples. However, it is insufficiently scientific to study these mechanical properties of rock under cyclic loading conditions in isolation, regardless of the coupling effect of internal high stress and external disturbance load.

Therefore, in an attempt to gain insight into understanding concerning how external dynamic disturbance loading, in terms of loading frequency and amplitude, coupling with high level of internal stress results in progressive damage and failure, a series of disturbance tests with different disturbance frequencies and amplitudes was carried out on sandstone samples at the axial stress level of 85% of uniaxial compressive strength. The main purpose of this chapter focuses on:

1. Investigating effect of frequency and amplitude of cyclic external disturbance load on stress–strain behaviors of rock samples.
2. Analyzing the progressive damage evolution of rock sample in the process of applying external disturbance loading, and comparing failure modes after that.

4.2 Testing Scheme and Procedure

The disturbance tests were all conducted under condition of uniaxial compression. As illustrated in Fig. 4.1a, loading process was divided into two phases: quasi-static loading phase and disturbance loading phase.

In the quasi-static loading phase of the test, the axial stress was continuously applied under axial displacement controlling mode with a rate of 0.02 mm/min up to a target set point which was the predetermined level (around 85%) of the uniaxial compressive strength (UCS). The reason why the 85% of UCS was chosen is because this axial stress level is relatively high for the rock samples but still in the linear elastic stage. By performing triaxial compressive tests on 3–5 rock samples for each level of confining pressure, the stress–strain curves (shown in Fig. 3.4) were obtained and the

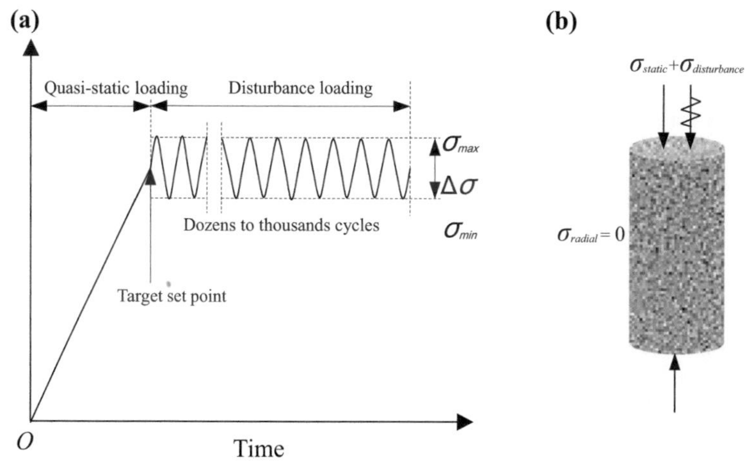

Fig. 4.1 Sketch map of loading method: **a** the loading path of experiment **b** stress state of sandstone sample under both disturbance loading and static loading

static uniaxial compressive strength was determined as 64.7 MPa. Thus, the target set point was specified as $64.7 \times 85\% \approx 55.0$ MPa in these disturbance loading tests.

In the disturbance loading phase, the axial stress was kept constant at the predetermined level of the target set point. In the meantime, the rod for disturbance loading was servo controlled to 'seat' on the rock sample by applying a small preload. The desired disturbance load with certain levels of frequency and amplitude was then apply on the top of the sample accurately with the close loop for disturbance loading. In this study, the disturbance loads were specified as sinusoidal waves. The disturbance frequencies of the waves were set as 5 Hz, 10 Hz, and 20 Hz, and the amplitudes were set as 1.0 MPa, 2.5 MPa, and 5.0 MPa, respectively. These values were chosen because they made the cyclic disturbance loads be slight waves with medium loading rate. It is noted that, as shown in Fig. 4.1a, disturbance amplitude herein refers to an absolute ($\pm$ value), whereby the value of amplitude is equal to one-half of the difference between the maximum stress and the minimum stress ($= \Delta\sigma/2$). The experimental conditions of each sample are listed in Table 4.1, and the stress state of the rock sample under both disturbance loading and static loading is shown in Fig. 4.1b.

The thermal effects were not taken into consideration in this study, and all of the tests were conducted under isothermal conditions at room temperature (controlled at ~ 22 °C with an accuracy of $\pm$ 1 °C).

4.3 Experimental Results and Discuss

4.3.1 Characteristics of Stress–Time Curves and Stress–Strain Curves

In the disturbance tests following the loading path shown in Fig. 4.1a, the stress–time curves (Fig. 4.2) were recorded for the whole loading duration with the disturbance frequencies of 5 Hz, 10 Hz, and 20 Hz and amplitudes of 1.0 MPa, 2.5 MPa, and 5 MPa, respectively. From the test results, the following phenomena can be observed: (1) Based on generally accepted division method for whole stress–strain curve of rock (Wang and Park 2002; Cai et al. 2004; Yang et al. 2012), the stress–time curves at the quasi-static loading phase obtained from the experiments mainly experience two different deformation stages, including crack closure stage and linear elastic stage. At the crack closure stage, the stress–time curves are slightly concave upward, which indicates that primary micro-pores and micro-fissures of the rock gradually get compacted under axial loading. The curves then show up as a constant slope for quite a long increasing stress duration at the linear elastic stage, in which the secondary cracks are steadily forming, growing, extending, propagating, and connecting under the continuously growing axial load. (2) At the disturbance loading phase, the frequency and amplitude of the axial disturbance stress have important effects on time duration of this phase. Specifically, with increased amplitude at a same

Table 4.1 Experimental conditions of each rock sample

Sample no.	Frequency/Hz	Amplitude/MPa	Sample no.	Frequency/Hz	Amplitude/MPa	Sample no.	Frequency/Hz	Amplitude/MPa
RS-1-1	5	1.0	RS-4-1	10	1.0	RS-7-1	20	1.0
RS-1-2	5	1.0	RS-4-2	10	1.0	RS-7-2	20	1.0
RS-1-3	5	1.0	RS-4-3	10	1.0	RS-7-3	20	1.0
RS-2-1	5	2.5	RS-5-1	10	2.5	RS-8-1	20	2.5
RS-2-2	5	2.5	RS-5-2	10	2.5	RS-8-2	20	2.5
RS-2-3	5	2.5	RS-5-3	10	2.5	RS-8-3	20	2.5
RS-3-1	5	5.0	RS-6-1	10	5.0	RS-9-1	20	5.0
RS-3-2	5	5.0	RS-6-2	10	5.0	RS-9-2	20	5.0
RS-3-3	5	5.0	RS-6-3	10	5.0	RS-9-3	20	5.0

frequency, the time duration of disturbance loading decreases. When the disturbance loading is applied at the same level of amplitude, the time duration also decreases with the increased frequency. (3) At failure of rock sample, the stress drops suddenly, indicating that the failure of the sandstone samples exhibits brittleness to some extent.

Figure 4.3 shows stress–strain curves—including axial strain curves and radial strain curves—corresponding to Fig. 4.2. It is found that, at the quasi-static loading phase, the stress increase with the axial strain in all cases changing from slightly concave upward shapes to the more-or-less straight lines. The radial strain (i.e., using negative signs as tensile strain) also increase at this phase. At the disturbance loading phase, even though the disturbance stresses are controlled and applied as regular sine curves on the rock samples, the corresponding stress–strain curves obtained are not actually quite regular. This is because the properties of rock materials are extremely dependent on the micro-pores and micro-fissures which make the rock samples exhibit their heterogeneity and nonlinearity. The heterogeneity and nonlinearity lead to the irregular cyclic stress–strain curves at the phase of disturbance loading, and the nonlinearity also makes the rock materials quite different from metal materials (Schijve 2009).

4.3.2 Effect of Disturbance Frequency and Amplitude on Mechanical Response of Rock Samples

Table 4.2 summarizes the number of cycles, the maximum axial strain and maximum radial strain before failure, and the volumetric strain when failure occurred for each sample. At the same disturbance frequency, the number of cycles decrease with the increase in disturbance amplitude; the higher the disturbance amplitude, the more cycles the rock sample can beard in the disturbance loading phase. The above phenomenon can be well explained from the perspective of strain energy: the strain energy of every single disturbance cycle is related to amplitude of stress but not to time, thus the strain energy input into the rock samples in every single disturbance cycle is identical even though the disturbance frequency is specified as different levels. If it is postulated that failure of the rock samples under disturbance loading would occur when the input energy reaches a certain value, the number of cycles that the rock sample experiences can be treated as the cumulative energy input into the rock sample, indicating that rock sample need more disturbance cycles to reach its critical failure condition with lower amplitudes. Besides, it is also can be observed from Table 4.2 that, at the same disturbance amplitude, the number of cycles also decline with the rise of disturbance frequency. In particular, in the condition of disturbance frequency of 20 Hz and amplitude of 2.5 MPa, the cycle number is even zero. That is to say, the rock sample cannot bear the disturbance load with a condition of relatively high frequency coupling with relatively high amplitude. If we still explain this phenomenon from the point of strain energy, it may lead us to a different conclusion that the number of cycles may not vary with the disturbance frequency because

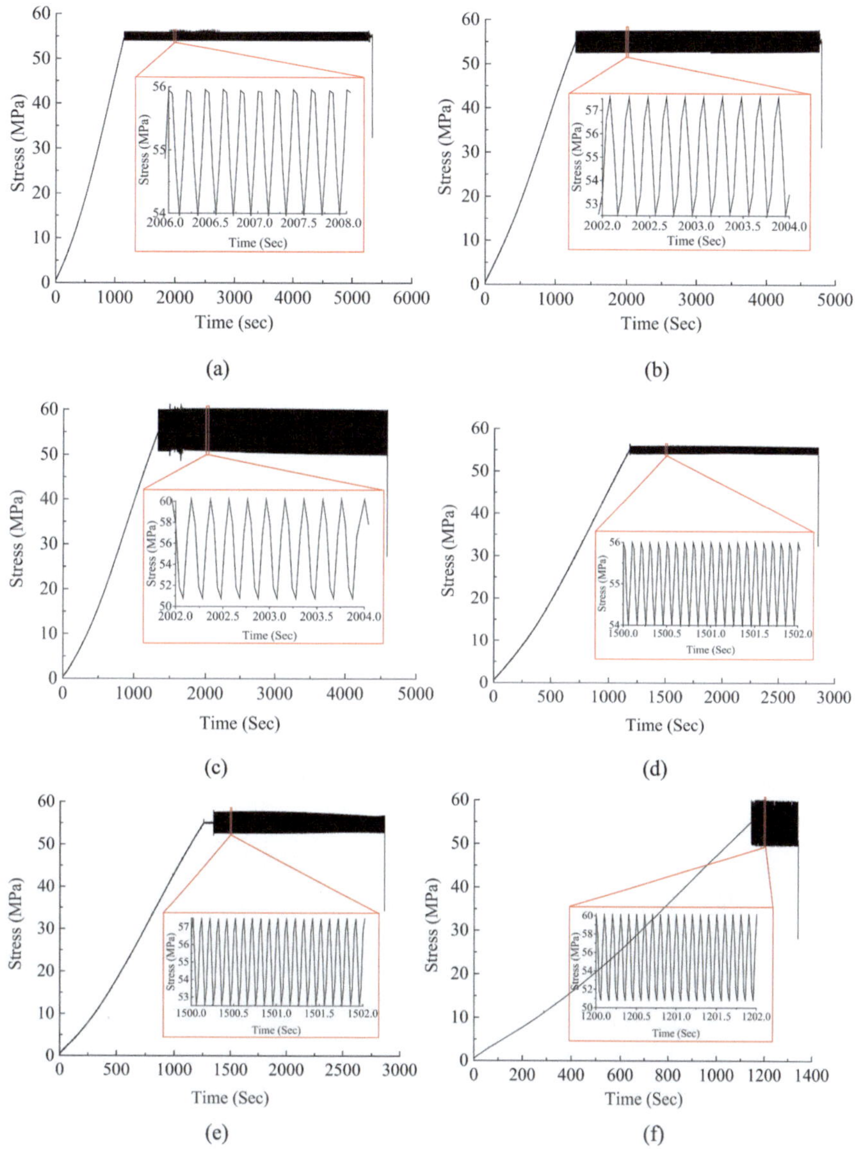

Fig. 4.2 Stress–time curves at coupling conditions of disturbance frequencies of 5 Hz, 10 Hz, and 20 Hz and amplitudes of 1.0 MPa, 2.5 MPa, and 5 MPa: **a** 1.0 MPa, 5 Hz, **b** 2.5 MPa, 5 Hz, **c** 5.0 MPa, 5 Hz, **d** 1.0 MPa, 10 Hz, **e** 2.5 MPa, 10 Hz, **f** 5.0 MPa, 10 Hz, **g** 1.0 MPa, 20 Hz, **h** 2.5 MPa, 20 Hz, **i** 5.0 MPa, 20 Hz

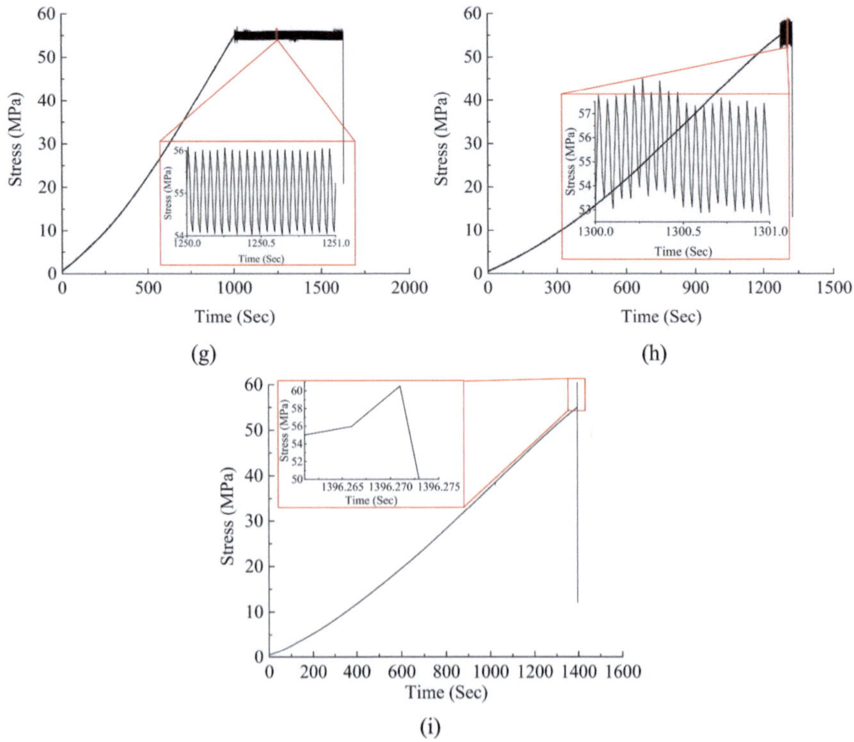

Fig. 4.2 (continued)

strain energy is time-invariant. However, Deng et al. (2017) reported a series of cyclic loading and unloading experiments on sandstone, and strongly verified that the elastic strain energy has little relation with the frequency of load in the cyclic loading and unloading process. According to their conclusion, the observation in this experiment that disturbance frequency has influence in time duration of the disturbance loading phase is reasonable and it can be mutually confirmed with the conclusion drawn by Deng et al. (2017). The reason for the influence may be that the disturbance frequencies of 5, 10, and 20 Hz is much larger than conventional cyclic loading and unloading compression tests, and the relatively higher frequencies lead to much bigger strain rates in disturbance loading tests than in conventional cyclic tests. This is the difference between low frequency cyclic loading and unloading tests and relatively high frequency disturbance loading tests, and also is the reason why the number of cycles decreases with the increase in disturbance frequency at a certain amplitude.

Comparing the maximum axial strains before failure in disturbance tests in Table 4.2 with the axial strain corresponding to peak stress in conventional uniaxial compression (0.5261%), it is found that these two axial strain values are very close with relatively errors less than $\pm$ 10% in all cases, indicating that maximum linear

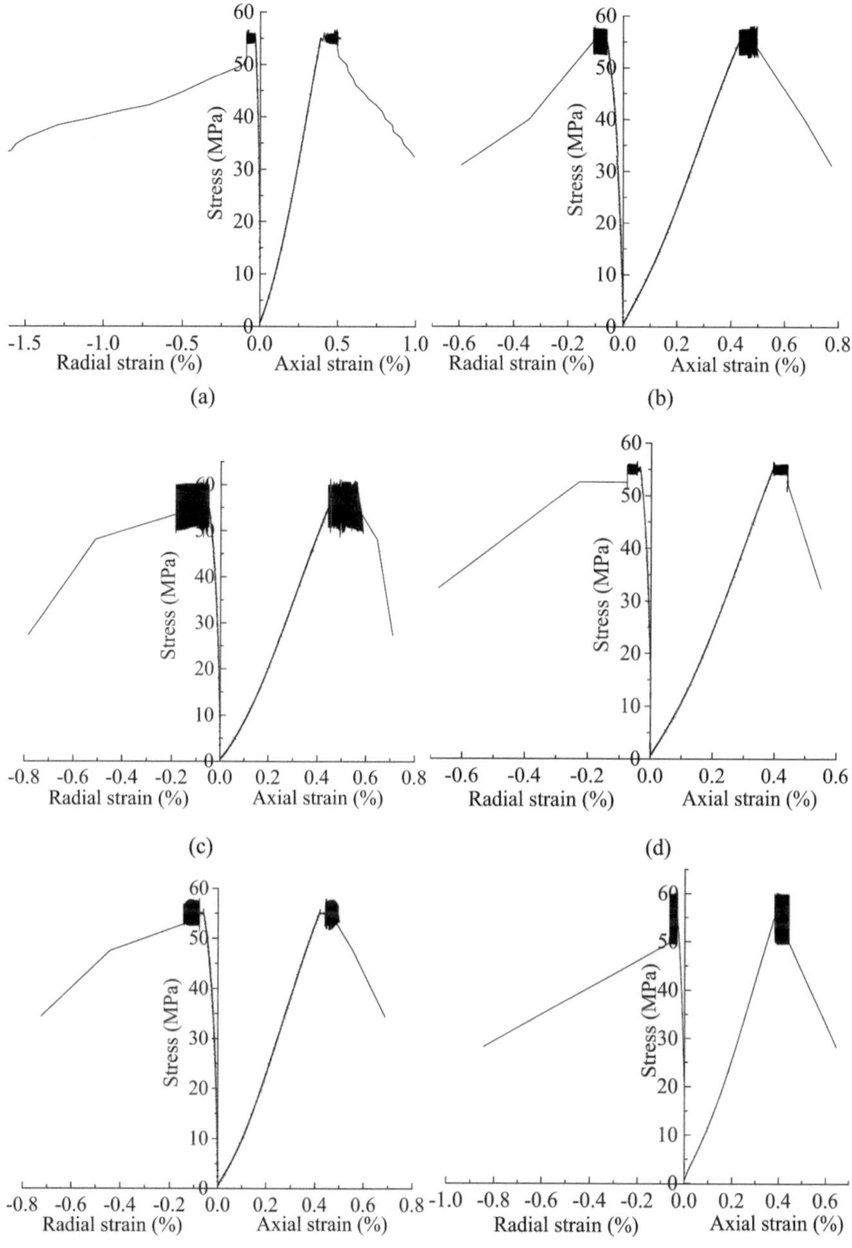

Fig. 4.3 Stress–strain curves at coupling conditions of disturbance frequencies of 5 Hz, 10 Hz, and 20 Hz and amplitudes of 1.0 MPa, 2.5 MPa, and 5 MPa: **a** 1.0 MPa, 5 Hz, **b** 2.5 MPa, 5 Hz, **c** 5.0 MPa, 5 Hz, **d** 1.0 MPa, 10 Hz, **e** 2.5 MPa, 10 Hz, **f** 5.0 MPa, 10 Hz, **g** 1.0 MPa, 20 Hz, **h** 2.5 MPa, 20 Hz, **i** 5.0 MPa, 20 Hz

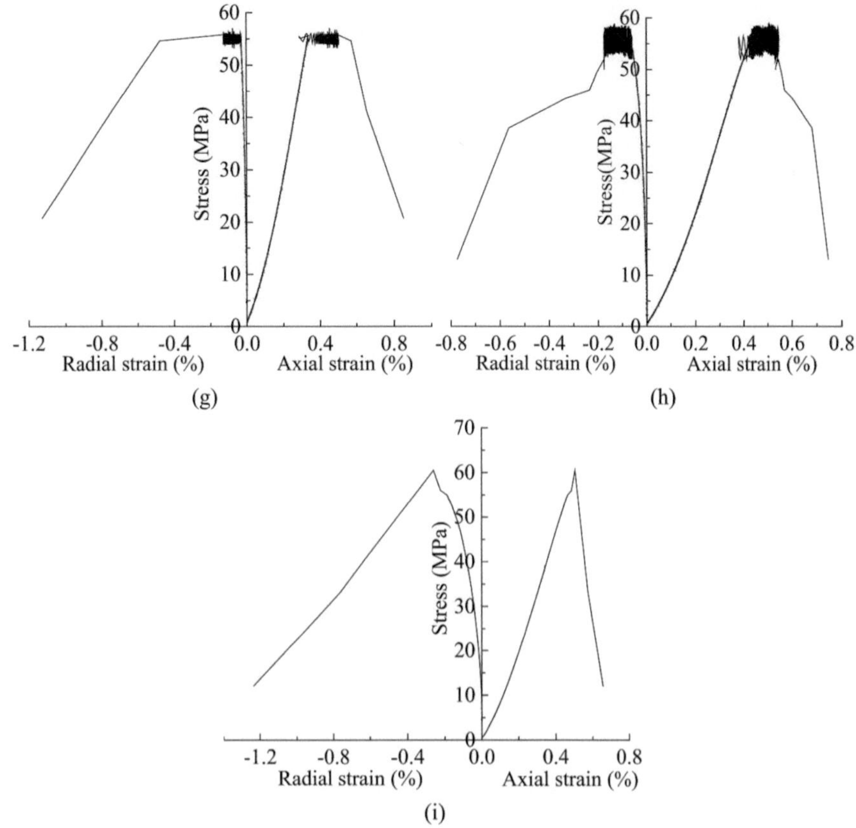

Fig. 4.3 (continued)

Table 4.2 Summary of the number of cycles, axial strain, radial strain and volumetric strain when failure occurred for each sample

Sample no.	Frequency/ Hz	Amplitude/ MPa	Number of cycles N	Maximum axial strain before failure/%	Maximum radial strain before failure/%	Volumetric strain at failure/%
RS-1-1	5	1.0	20,883	0.4960	0.0898	0.3132
RS-2-2	5	2.5	18,861	0.4967	0.1080	0.2642
RS-3-1	5	5.0	16,261	0.5871	0.1850	0.1942
RS-4-3	10	1.0	16,752	0.4765	0.0775	0.3178
RS-5-3	10	2.5	15,180	0.5001	0.1442	0.2037
RS-6-2	10	5.0	3987	0.4425	0.0643	0.3138
RS-7-1	20	1.0	12,473	0.5058	0.1257	0.2492
RS-8-3	20	2.5	1009	0.5496	0.1739	0.1886
RS-9-2	20	5.0	0	0.4650	0.1884	0.0882

strain theory can be a good failure criterion for disturbance tests. The maximum radial strains before failure in disturbance tests are all much less than the radial strain corresponding to peak stress in conventional uniaxial compression (0.4592%)—i.e., the growths of radial strain under disturbance loading conditions are far less than under conventional uniaxial compressive condition. Correspondingly, the values of volumetric strains at failure listed in Table 4.2 are positive in all cases, which means the rock samples at failure in disturbance loading tests are in a volume compressed state relative to the initial volume. The dilation of the rock samples occurs after the rapid stresses drop, making this phenomenon be quite different from uniaxial compression tests, in which the failure of rock samples caused by dilatancy (Brace et al. 1966).

In the disturbance loading phase of the tests following the load path show in Fig. 4.1, the variation of axial and radial strain with time were recorded from the beginning of disturbance load applying to the failure of rock samples. Figure 4.4 presents the experimental results with applied frequencies of 5 Hz, 10 Hz, and 20 Hz, respectively. From the results, the strains increase quickly with time at the beginning of the disturbance loading phase, and subsequently increase gently. These changing trends are similar with regime I and regime II—transient creep and steady-state creep—in idealized one-dimensional creep curve (Jaeger et al. 2007). Similar observations were made by Akai and Ohnishi (1983) in their cyclic loading experiments on marble and sandstone specimens. The reason for these trends is that, after the stress reaches to and stabilizes at the target set point in quasi-static loading phase, the axial strain in disturbance loading phase is a combination of two parts: One part is caused by creep and the other is caused by pure disturbance. The creep triggered part makes the strain–time curve similar with the first two regimes of creep curve; while the pure disturbance triggered part makes their differences—there is no regime III of tertiary creep in strain–time curves. Besides, it is also can be observed from Fig. 4.4 that the increment of the axial and radial strains grows with the disturbance amplitude with a same certain frequency; i.e., the larger the amplitude, the more obvious effect of disturbance loading on axial and radial strain of rock sample.

4.3.3 Damage Evolution of Rock Samples During Disturbance

In solid mechanics, damage is employed to describe the discontinuous characteristics of solid materials, such as initializing, forming, growing, propagating and connecting of the inner micro-pores and micro-cracks, but the solid materials are still considered to be continuous in macroscopic scale (Zhu et al. 2015a, b, 2016; Rabczuk and Ren 2017). Damage evolution always accompanies by degradation of many physical and mechanical properties of materials, such as elastic modulus, stiffness (Areias et al. 2016, 2018), ultrasonic velocity, residual strength, electrical resistivity, acoustic emission counts, compliance (Ren et al. 2016, 2017), damaged volume (Rabczuk and Belytschko 2004; Rabczuk et al. 2010), number of bonds (Zhou et al. 2017), etc. Thus,

Fig. 4.4 Variation of axial and radial strains with time under the three levels of frequencies in the disturbance loading phase (the zero points of time corresponds to start of disturbance load)

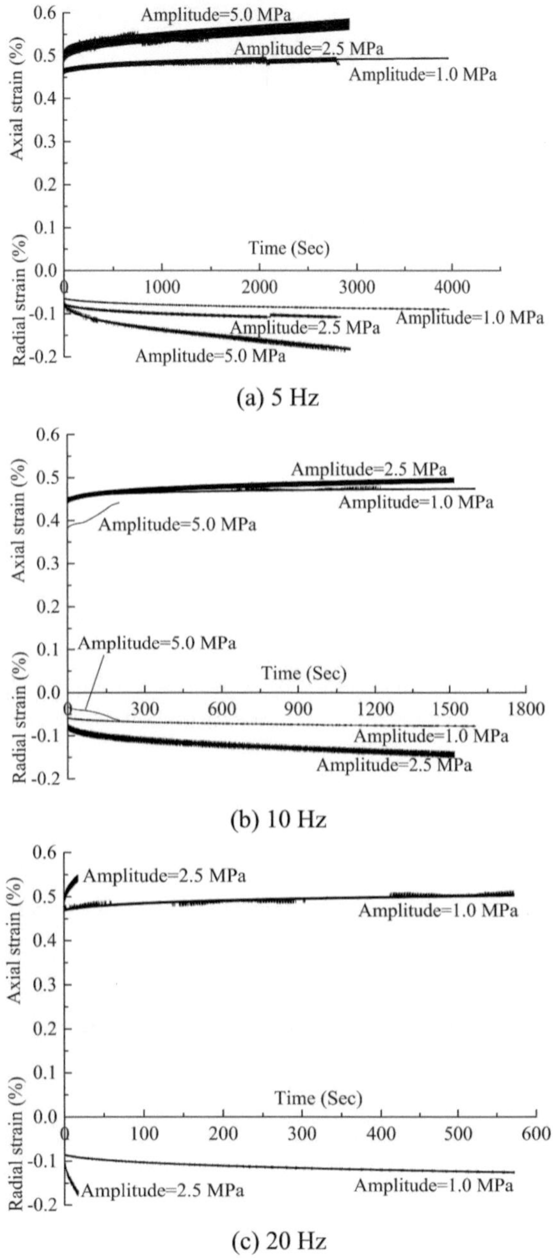

damage variable can not only be physically defined by the surface density of micro-cracks and intersections of micro-voids based on the cross section (Lemaitre and Desmorat 2005), but also can be described with the ratio of pre- to post-degradation of those physical and mechanical properties. No matter which parameter is chosen to describe the damage process during disturbance loading phase, damage evolution is supposed to be in compliance with propagation of micro-cracks in the rock samples. In order to get certainty, Xiao et al. (2010) summarized four basic requirements that the damage variable defined must meet: (1) It has clear physical significance, (2) it can be measured easily and applied in engineering conveniently, (3) its evolution law coincides well with the actual degradation process of material, and (4) it can take the initial damage into account. To meet the abovementioned four requirements, four different methods are used to define damage variables as follow.

(1) Damage variables based on residual strain method

Liu and He (2012) defined residual strain as the axial strain at which the axial load reaches the minimal value in a cycle after *n* cycles, and proposed the definition of damage variable based residual strain method as:

$$D = \frac{\varepsilon_r^n}{\varepsilon_r^f} \tag{4.1}$$

where ε_r^n and ε_r^f are the residual axial strain after n cycles and the ultimate residual axial strain at failure in cyclic loading and unloading tests, respectively. By introducing this damage variable into the disturbance loading tests, Fig. 4.5 presents the relationship between residual strain and number of disturbance cycles for each rock sample, which is B-spline fitted. For all these cases, as the number of cycles increases, the residual strain increases rapidly first and then slowly. Figure 4.6 illustrates relationship between damage variable *D* and relative cycles *n*/*N*. This relationship curves can well reflect the gradual increase in damage degree of rock samples in disturbance loading phase under different disturbance frequencies and different amplitudes conditions. The first point of each curve in Fig. 4.6 also characterizes the damage degree after quasi-static loading phase, i.e. the initial damage degree of disturbance loading phase. In addition, it can also be concluded from Fig. 4.6 that, when subjected to disturbance loading, the evolution of damage variable exhibits same trend of residual strain, i.e., increase sharply first and subsequently increase slowly. And the growth rate of damage variable that increase from initial damage to complete damage ($D = 1$) is more modestly with a relatively larger coupling condition of amplitude and frequency of disturbance load.

(2) Damage variables based on average dynamic axial stiffness method

As shown in Fig. 4.7, Liu et al. (2012) found that the typical deviatoric stress-axial strain curve of intact rock samples under triaxial cyclic loading and unloading has a significant feature: In the early stage of cyclic loading and unloading, the axial strain developed in one cycle is small, the distribution of the adjacent two cycles is dense, and the difference between the axial strain corresponding to the wave crest and the axial strain corresponding to the wave trough is small. With the continuous cyclic loading and unloading, the axial strain developed by the rock sample in one cycle gradually becomes larger, and the distribution of the adjacent two cycle periods becomes sparse, and the difference between the axial strain corresponding to the wave crest and the axial strain corresponding to the wave trough gradually increases.

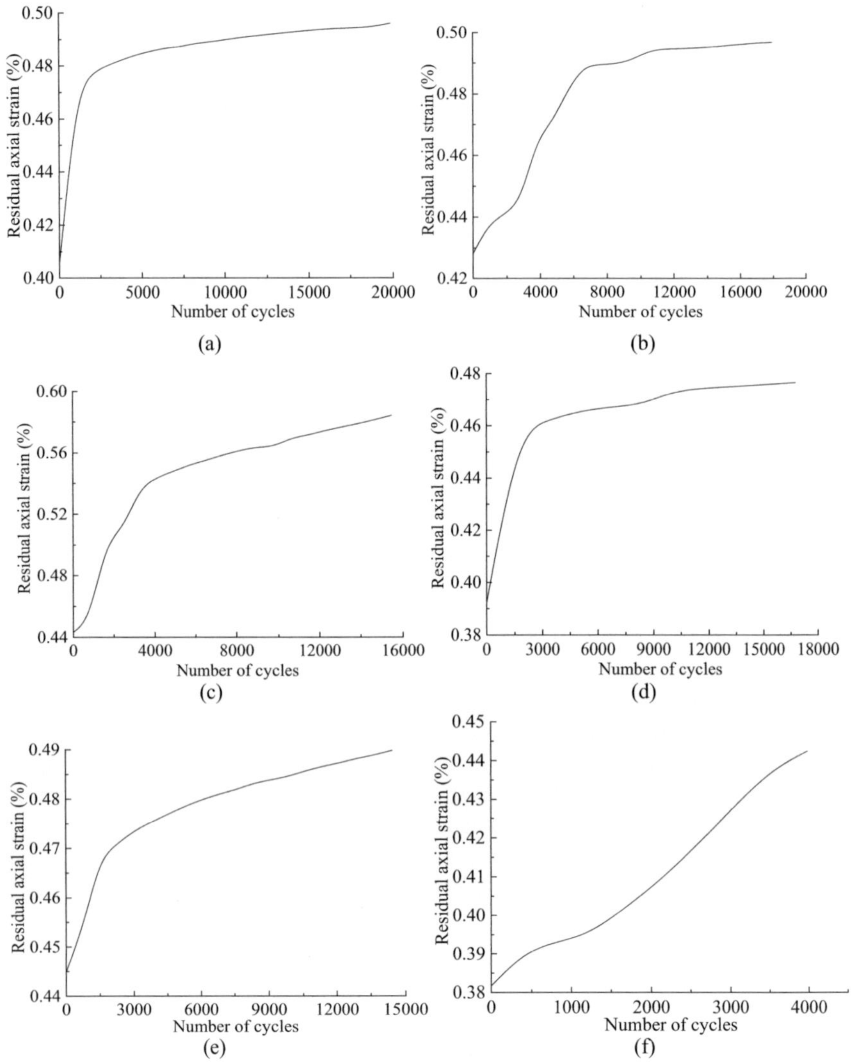

Fig. 4.5 Relationship between residual strain and number of disturbance cycles: **a** 1.0 MPa, 5 Hz, **b** 2.5 MPa, 5 Hz, **c** 5.0 MPa, 5 Hz, **d** 1.0 MPa, 10 Hz, **e** 2.5 MPa, 10 Hz, **f** 5.0 MPa, 10 Hz, **g** 1.0 MPa, 20 Hz, **h** 2.5 MPa, 20 Hz

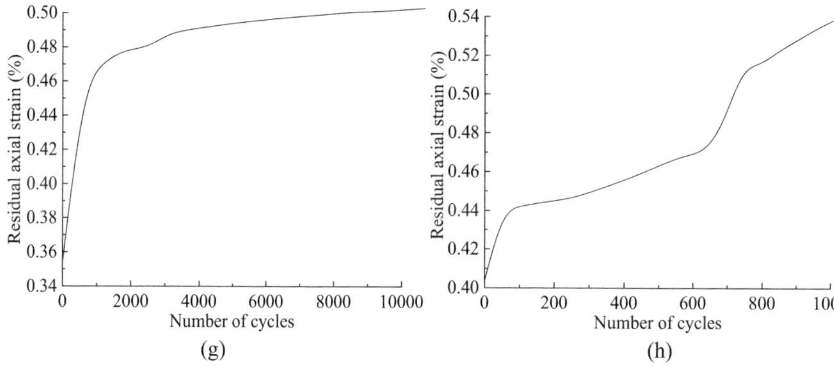

Fig. 4.5 (continued)

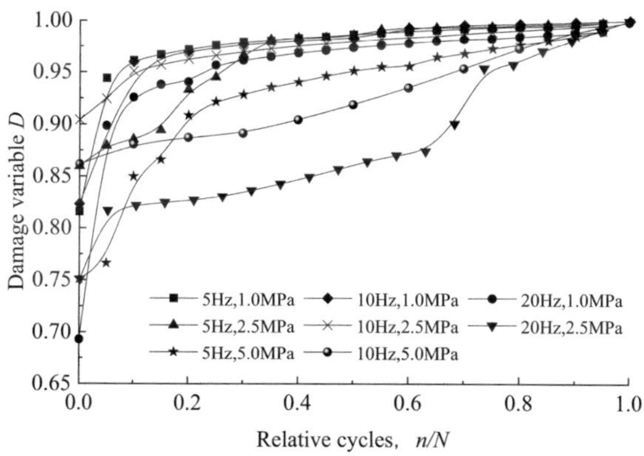

Fig. 4.6 Relationship between damage variable D and relative cycles n/N

This phenomenon can also indirectly indicate that the axial stiffness of the rock is constantly changing under the action of cyclic disturbance, showing a trend of getting smaller and smaller. Based on this phenomenon, the concept of "average dynamic axial stiffness" can be defined.

Figure 4.8 gives a schematic representation of the method of defining the average dynamic axial stiffness. For any disturbance cycle in the deviatoric stress-axial strain curve obtained from a cyclic loading and unloading test, the stress difference between the peak and the trough is Δ_{stress}, and the axial strain difference between the peak and the trough is Δ_{strain}. Then the average dynamic axial stiffness is defined as (Liu et al. 2012)

$$A_{sd} = \frac{\Delta_{stress}}{\Delta_{strain}}. \tag{4.2}$$

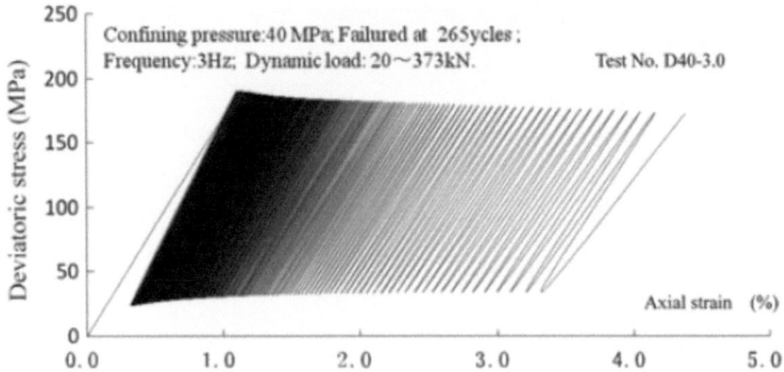

Fig. 4.7 Typical deviatoric stress–strain of a sandstone sample under triaxial cyclic loading and unloading condition (Liu et al. 2012)

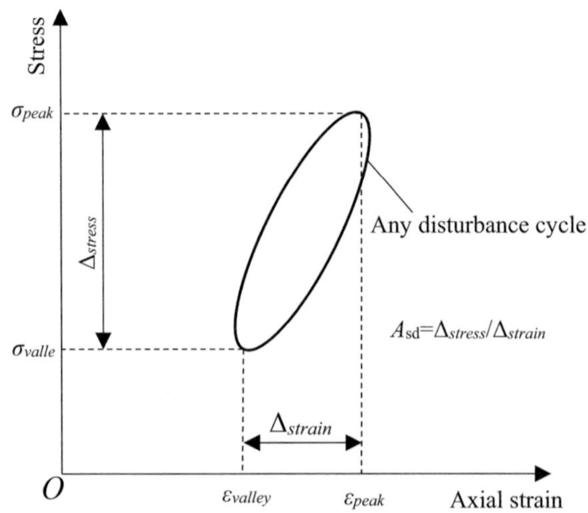

Fig. 4.8 Schematic diagram of the method to define average dynamic axial stiffness

In the disturbance loading stage of the cyclic dynamic disturbance test, the Δ_{stress} in Eq. (4.2) is actually twice the disturbance amplitude (corresponding to $\Delta\sigma$ in Fig. 4.1a), which in this test is 2.0 MPa, 5.0 MPa, and 10.0 MPa, respectively. Δ_{strain} can be calculated from the original stress–strain data obtained from the test. Due to the large number of disturbance cycles of most rock samples in this test and the abnormal phenomenon of sudden jump of some data points in the test, the axial strain difference Δ_{strain} corresponding to the peak and the trough of 20 cycles was uniformly extracted from each stress–strain curve. By substituting the 20 sets of data for each typical rock sample into Eq. (4.2), the average dynamic axial stiffness values corresponding to the 20 cycles of each typical rock sample can be obtained, which are listed in Table 4.3.

As shown in Fig. 4.9, the variation trends of average dynamic axial stiffness with increasing number of disturbance cycles were plotted for each typical rock specimen using B-spline fitting based on the data in Table 4.3. The results demonstrate that for most specimens, regardless of whether the amplitude and frequency were set at relatively high or low levels, the average dynamic axial stiffness gradually

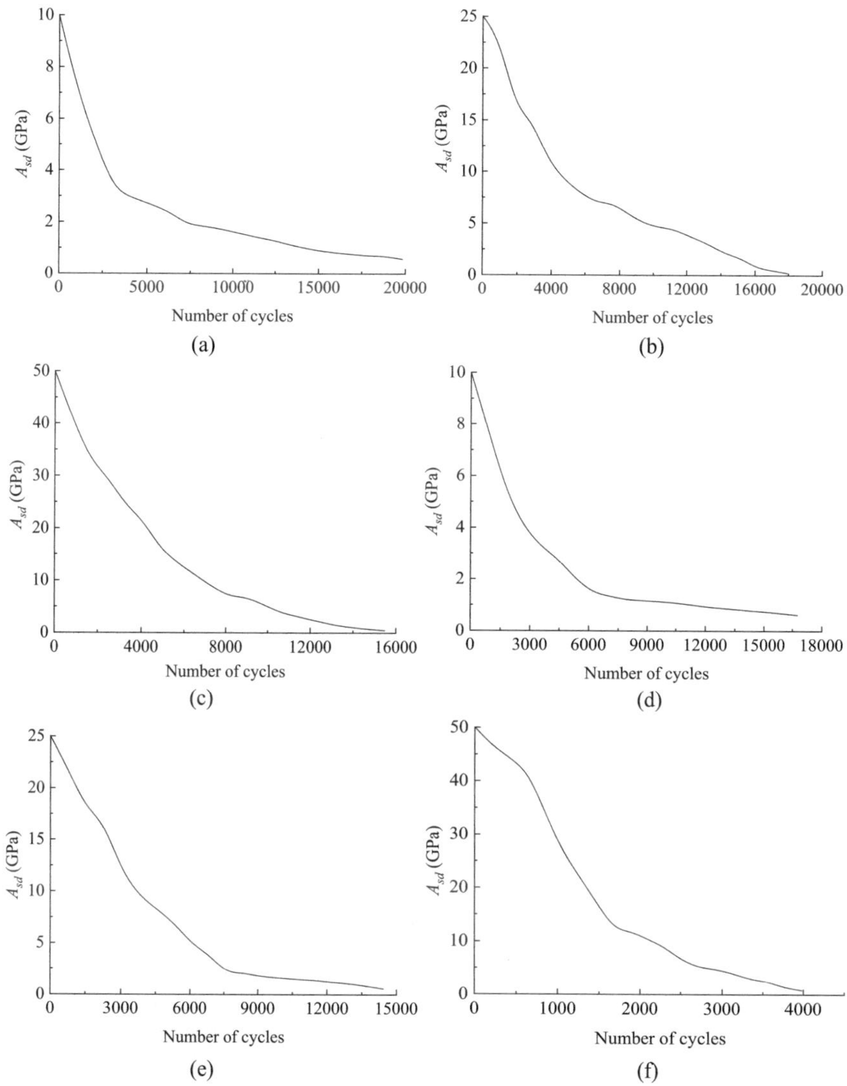

Fig. 4.9 Relationship between average dynamic axial stiffness and number of cyclic dynamic disturbance cycles: **a** 1.0 MPa, 5 Hz, **b** 2.5 MPa, 5 Hz, **c** 5.0 MPa, 5 Hz, **d** 1.0 MPa, 10 Hz, **e** 2.5 MPa, 10 Hz, **f** 5.0 MPa, 10 Hz, **g** 1.0 MPa, 20 Hz, **h** 2.5 MPa, 20 H z

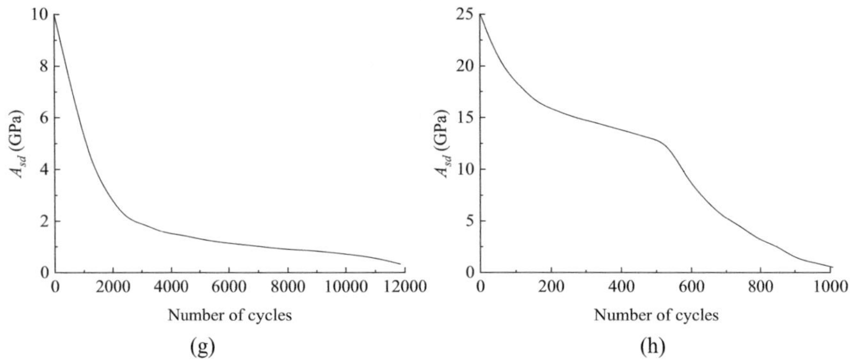

Fig. 4.9 (continued)

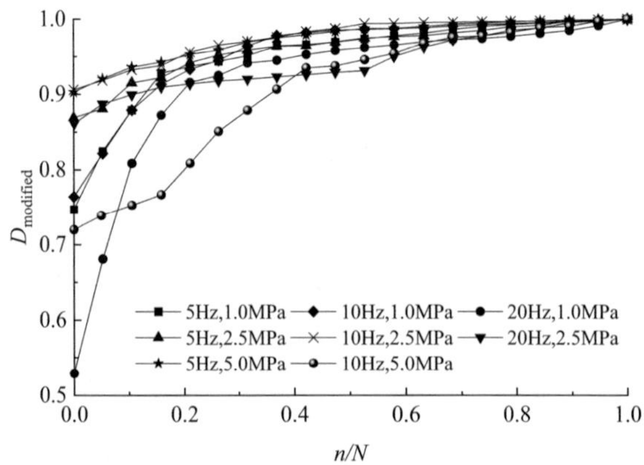

Fig. 4.10 Relationship between damage variable $D_{modified}$ and relative cycles n/N based on the method of average dynamic axial stiffness

decreased with progressive disturbance cycles - showing a relatively rapid decline during the initial loading stage followed by a more gradual reduction in the middle and later stages. The only exception was specimen RS-9-2, which failed instantly upon application of the cyclic micro-disturbance load.

As previously mentioned, under sustained cyclic disturbance loading, the sandstone specimens exhibited characteristic stress-strain curves featuring progressively accelerated axial strain development coupled with gradually decreasing average dynamic axial stiffness as disturbances continued. Building upon these characteristics, we can further define a damage variable based on the average dynamic axial stiffness during cyclic micro-disturbance (Liu and He, 2012):

$$D = \frac{A_{sd,max} - A_{sd}}{A_{sd,max} - A_{sd,min}} \tag{4.3}$$

where $A_{sd,max}$ represents the maximum average dynamic axial stiffness, typically corresponding to the first disturbance cycle, $A_{sd,min}$ denotes the minimum average dynamic axial stiffness, generally corresponding to the last disturbance cycle before failure under cyclic micro-disturbance, A_{sd} indicates the average dynamic axial stiffness during any arbitrary cycle in the dynamic disturbance process.

As illustrated in Fig. 4.9, the rock's average dynamic axial stiffness (A_{sd}) reaches its maximum value ($A_{sd} = A_{sd,max}$) at the initiation of cyclic dynamic disturbance. According to Eq. (4.3), this results in a damage variable D of 0 at the starting point, which contradicts the fourth fundamental requirement for damage variable definition as proposed by Xiao et al. (2010). Specifically, the damage variable defined by the average dynamic axial stiffness method fails to account for initial damage. To incorporate the initial damage condition existing at the commencement of cyclic micro-disturbance loading (i.e., at the end of the quasi-static loading phase), Eq. (4.3) has been modified as follows:

$$D_{modified} = D_0 + \frac{A_{sd,max} - A_{sd}}{A_{sd,max} - A_{sd,min}} \cdot (1 - D_0) \tag{4.4}$$

where $D_{modified}$ represents the damage variable during cyclic dynamic disturbance based on the modified average dynamic axial stiffness method, 4_0 denotes the initial damage degree at the commencement of cyclic micro-disturbance loading (i.e., at the end of the quasi-static loading phase). Its value can be determined using test data from the quasi-static loading phase through other damage variable calculation methods. For this experiment, the calculated values of D_0 are presented in Table 4.4.

Using the average dynamic axial stiffness data (A_{sd}) of each rock specimen from Table 4.3 and the initial damage data (D_0) at the onset of cyclic dynamic disturbance loading from Table 4.4, the evolution data of damage variables during the cyclic micro-disturbance loading process can be calculated according to Eq. (4.4). Consequently, as shown in Fig. 4.10, the relationship curves between the modified damage variable $D_{modified}$ (based on the average dynamic axial stiffness method) and the relative cycle number n/N can be plotted.

As evidenced in Fig. 4.10, the modified damage variable based on the average dynamic axial stiffness method effectively captures both: (1) the initial damage state of rock specimens at the commencement of cyclic dynamic disturbance loading, and (2) the progressive damage degradation process during cyclic dynamic disturbance under uniaxial high-stress conditions. The average dynamic axial stiffness, as a physically meaningful parameter that can be directly measured during cyclic dynamic disturbance tests, provides a solid foundation for this modified damage variable. Consequently, this methodology proves to be theoretically sound and practically viable for quantifying rock damage evolution induced by cyclic dynamic disturbance in high-stress environments.

(3) Damage variables based on maximum axial and radial strain method

As previously described, during the cyclic dynamic disturbance tests, the axial strain developed within a single cycle is relatively small during the initial loading phase,

Table 4.3 Number of disturbance cycles and average dynamic axial stiffness (incomplete statistics)

RS-1-1 (1.0 MPa, 5 Hz)		RS-2-2 (2.5 MPa, 5 Hz)		RS-3-1 (5.0 MPa, 5 Hz)		RS-4-3 (1.0 MPa, 10 Hz)	
Number of disturbance cycles	A_{sd}	Number of disturbance cycles	A_{sd}	Number of disturbance cycles	A_{sd}	Number of disturbance cycles	A_{sd}
1	10	1	25	1	50	1	10
1044	7.11	943	22.76	814	41.09	882	7.71
2088	5.08	1886	16.26	1628	33.17	1763	5.43
3132	3.25	2829	14.98	2442	29.72	2645	4.07
4176	2.92	3772	11.20	3256	24.48	3527	3.28
5220	2.67	4715	9.28	4070	21.41	4408	2.81
6264	2.39	5658	8.01	4884	16.06	5290	2.04
7308	1.90	6601	7.03	5698	13.29	6172	1.49
8352	1.83	7544	6.96	6512	11.04	7053	1.34
9396	1.71	8487	5.94	7326	8.83	7935	1.20
10,440	1.55	9430	5.02	8140	6.95	8817	1.17
11,484	1.40	10,373	4.66	8954	6.82	9698	1.13
12,528	1.27	11,316	4.41	9768	5.44	10,580	1.08
13,572	1.08	12,259	3.70	10,582	3.90	11,462	0.99
14,616	0.94	13,202	3.09	11,396	3.11	12,345	0.91
15,660	0.83	14,145	2.17	12,210	2.36	13,225	0.86
16,704	0.77	15,088	1.74	13,024	1.57	14,107	0.80
17,748	0.70	16,031	0.79	13,838	1.08	14,989	0.75
18,792	0.69	16,974	0.48	14,652	0.74	15,871	0.69
19,836	0.57	17,917	0.24	15,466	0.48	16,752	0.62
RS-5-3 (1.0 MPa, 5 Hz)		RS-6-2 (2.5 MPa, 5 Hz)		RS-7-1 (5.0 MPa, 5 Hz)		RS-8-3 (1.0 MPa, 10 Hz)	
Number of disturbance cycles	A_{sd}	Number of disturbance cycles	A_{sd}	Number of disturbance cycles	A_{sd}	Number of disturbance cycles	A_{sd}
1	25	1	50	1	10	1	25
759	21.77	199	46.7	623	6.88	53	20.38
1518	18.09	419	44.24	1246	4.26	106	18.23
2277	16.68	628	41.79	1869	2.95	159	16.44
3036	11.91	838	34.41	2492	2.06	212	15.71
3795	9.55	1047	27.04	3115	1.87	265	15.02
4554	8.26	1257	22.12	3738	1.53	318	14.6
5313	6.88	1466	17.21	4361	1.46	371	14.08
6072	4.92	1676	12.29	4984	1.3	424	13.59

(continued)

Table 4.3 (continued)

RS-5-3 (1.0 MPa, 5 Hz)		RS-6-2 (2.5 MPa, 5 Hz)		RS-7-1 (5.0 MPa, 5 Hz)		RS-8-3 (1.0 MPa, 10 Hz)	
Number of disturbance cycles	A_{sd}	Number of disturbance cycles	A_{sd}	Number of disturbance cycles	A_{sd}	Number of disturbance cycles	A_{sd}
6831	3.87	1885	11.83	5607	1.19	477	13.04
7590	2.12	2095	10.37	6230	1.11	530	12.63
8349	2.1	2304	8.92	6853	1.04	583	9.33
9108	1.75	2514	6.46	7476	0.95	636	7.15
9867	1.62	2723	5.09	8099	0.89	689	5.45
10,626	1.48	2933	4.77	8722	0.86	742	4.48
11,385	1.39	3142	3.92	9345	0.8	795	3.16
12,144	1.21	3352	2.85	9968	0.72	848	2.59
12,903	1.09	3561	2.46	10,591	0.64	901	1.36
13,662	0.84	3771	1.32	11,214	0.51	954	0.97
14,421	0.58	3980	0.84	11,837	0.33	1007	0.53

Table 4.4 The values of damage variables obtained from the theoretical model at the end of quasi-static loading stage for each typical rock sample

Sample no.	Frequency/Hz	Amplitude/MPa	Axial strain at the end of quasi-static loading phase	Theoretical value of initial damage variable D_0
RS-1-1	5	1.0	0.3876	0.747
RS-2-2	5	2.5	0.4282	0.869
RS-3-1	5	5.0	0.4432	0.903
RS-4-3	10	1.0	0.3921	0.764
RS-5-3	10	2.5	0.4448	0.906
RS-6-2	10	5.0	0.3802	0.720
RS-7-1	20	1.0	0.3350	0.529
RS-8-3	20	2.5	0.4249	0.861
RS-9-2	20	5.0	0.4650	0.942

with closely spaced adjacent cycles. With continued cyclic loading-unloading disturbance, the axial strain per cycle progressively increases, resulting in increasingly dispersed spacing between adjacent cycles. Based on this characteristic, the damage variable can alternatively be defined using the maximum strain per cycle to characterize the progressive damage degradation of rock specimens until failure during cyclic micro-disturbance testing. The specific definition is given by the following equation (Xiao et al. 2010):

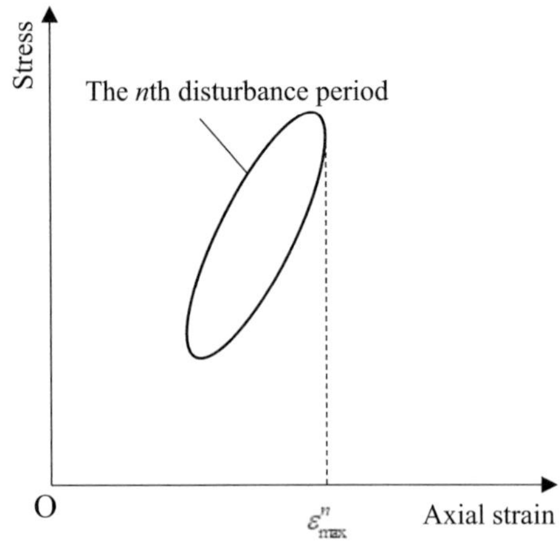

Fig. 4.11 Schematic diagram of the method to define the maximum axial strain

$$D = \frac{\varepsilon^n_{max} - \varepsilon^0_{max}}{\varepsilon^f_{max} - \varepsilon^0_{max}} \tag{4.5}$$

where ε^0_{max} and ε^f_{max} correspond to the maximum strain values during the initial loading cycle and the failure cycle when the specimen ruptures under cyclic dynamic disturbance, respectively. ε^n_{max}, as illustrated in Fig. 4.11, is the *n*th loading cycle of dynamic disturbance.

Similar to the damage variable defined by the average dynamic axial stiffness method (Eq. (4.3)), when cyclic dynamic disturbance loading is initially applied, the maximum strain at the *n*th loading cycle ε^n_{max} equals to maximum strain values at the initial loading cycle ε^0_{max}. Consequently, the damage variable *D* in Eq. (4.5) equals 0 at the starting point of cyclic dynamic disturbance. Likewise, the damage variable defined using the maximum strain method fails to account for initial damage. Therefore, Eq. (4.5) must be modified to obtain the revised damage variable based on the maximum strain method:

$$D_{modified} = D_0 + \frac{\varepsilon^n_{max} - \varepsilon^0_{max}}{\varepsilon^f_{max} - \varepsilon^0_{max}} \cdot 1 - D_0 \tag{4.6}$$

where D_0 represents the initial damage degree of the rock specimen at the commencement of cyclic dynamic disturbance loading (i.e., at the end of the quasi-static loading phase), with specific numerical values provided in Table 4.4.

Since the current experiment simultaneously acquired both axial strain and radial strain data, the modified damage variables based on the maximum axial strain method and maximum radial strain method can be analyzed separately. Although the volumetric strain could theoretically be calculated from the axial and radial strain data using elastic mechanics theory—allowing for the definition of a modified damage

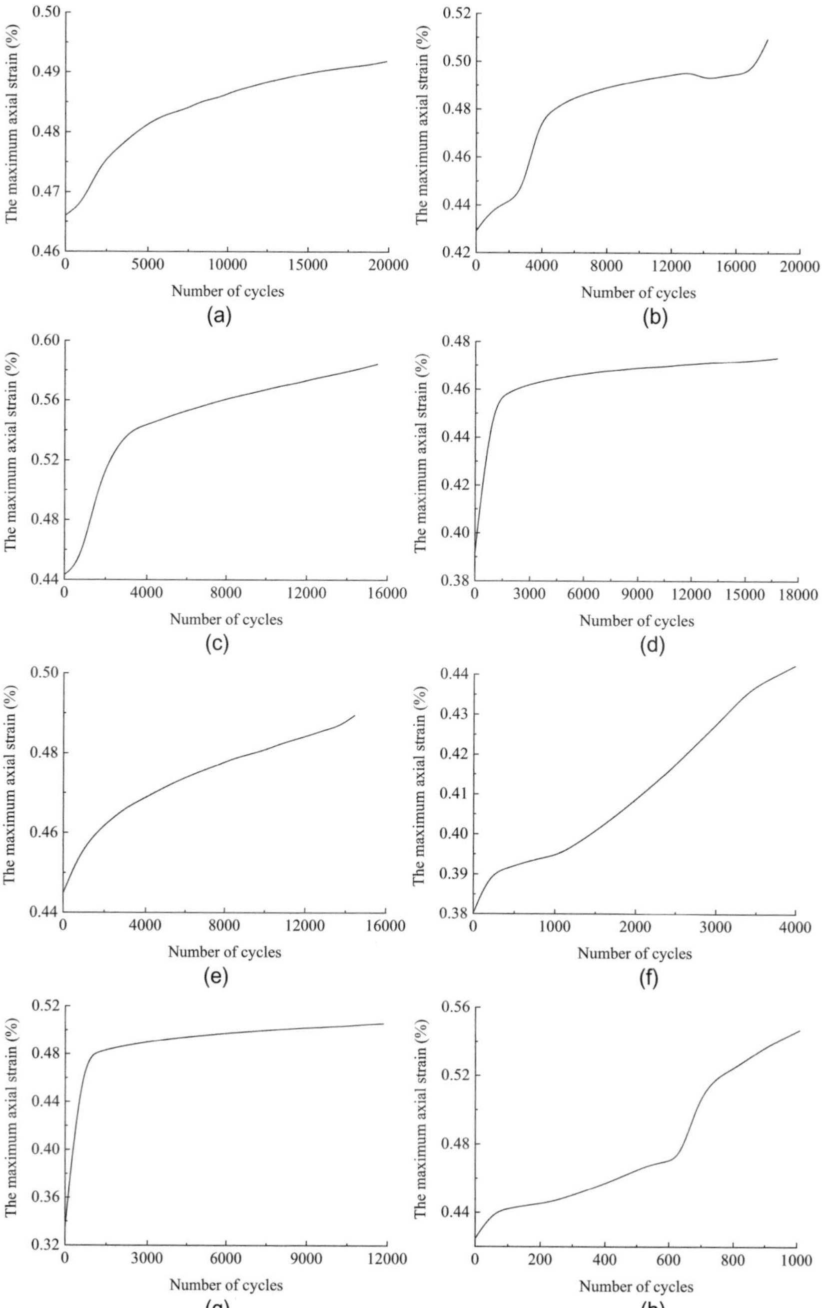

Fig. 4.12 Relationship between maximum axial strain and number of cyclic dynamic disturbance cycles: **a** 1.0 MPa, 5 Hz, **b** 2.5 MPa, 5 Hz, **c** 5.0 MPa, 5 Hz, **d** 1.0 MPa, 10 Hz, **e** 2.5 MPa, 10 Hz, **f** 5.0 MPa, 10 Hz, **g** 1.0 MPa, 20 Hz, **h** 2.5 MPa, 20 Hz

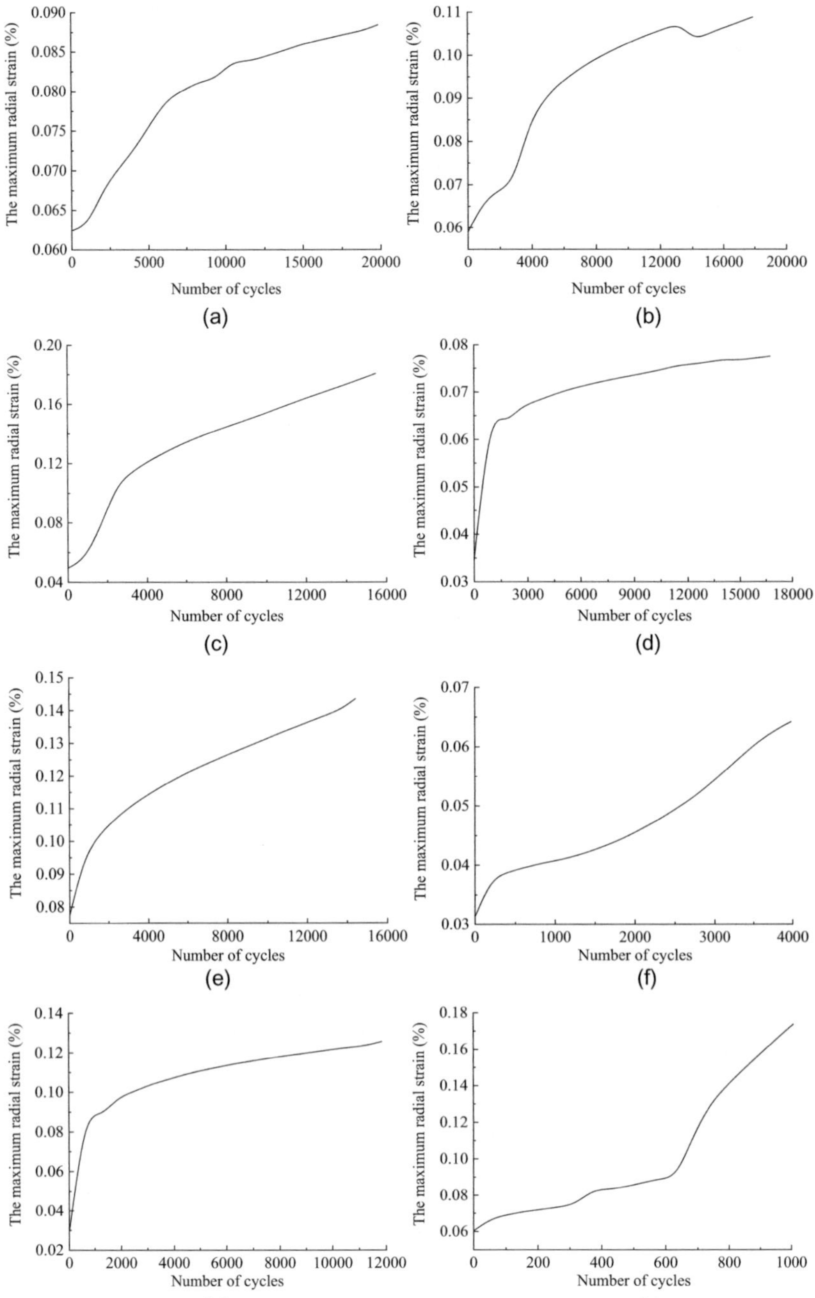

Fig. 4.13 Relationship between maximum radial strain and number of cyclic dynamic disturbance cycles: **a** 1.0 MPa, 5 Hz, **b** 2.5 MPa, 5 Hz, **c** 5.0 MPa, 5 Hz, **d** 1.0 MPa, 10 Hz, **e** 2.5 MPa, 10 Hz, **f** 5.0 MPa, 10 Hz, **g** 1.0 MPa, 20 Hz, **h** 2.5 MPa, 20 Hz

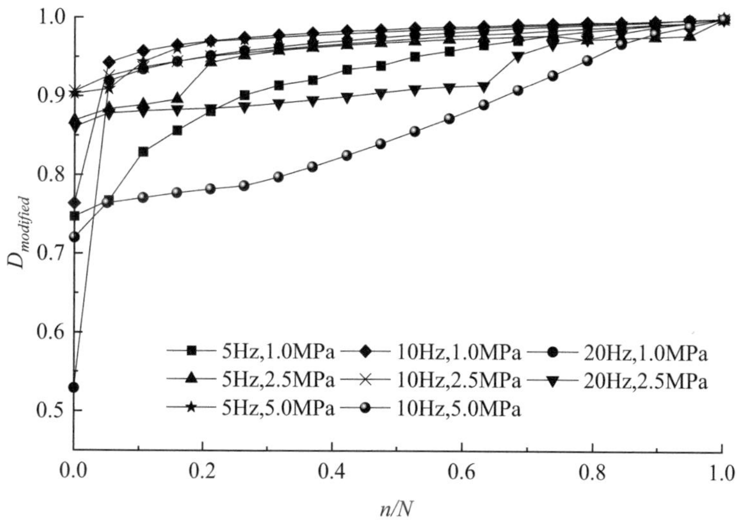

(a) Maximum axial strain method

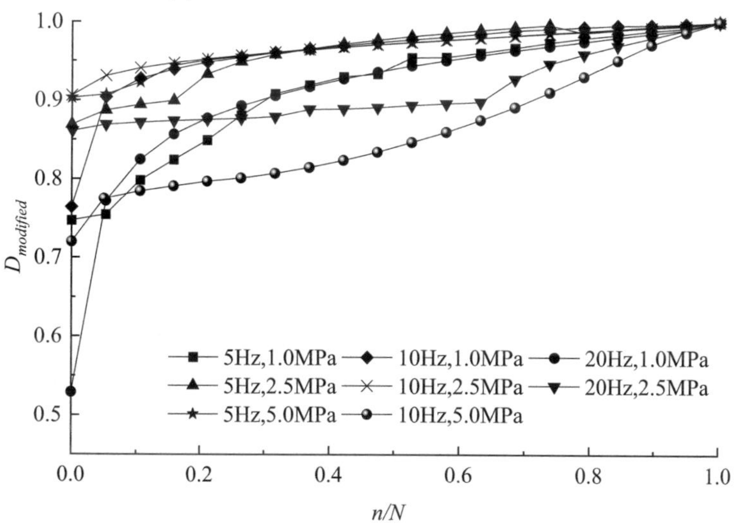

(b) Maximum radial strain method

Fig. 4.14 Relationship between damage variable $D_{modified}$ and relative cycles n/N based on maximum axial and radial strain method

Table 4.5 Number of disturbance cycles and maximum axial strain for the typical rock samples (incomplete statistics)

RS-1-1 (1.0 MPa, 5 Hz)		RS-2-2 (2.5 MPa, 5 Hz)		RS-3-1 (5.0 MPa, 5 Hz)		RS-4-3 (1.0 MPa, 10 Hz)	
Number of disturbance cycles	Maximum axial strain	Number of disturbance cycles	Maximum axial strain	Number of disturbance cycles	Maximum axial strain	Number of disturbance cycles	Maximum axial strain
1	0.4660	1	0.4290	1	0.4433	1	0.3921
1044	0.4681	943	0.4380	814	0.4521	882	0.4536
2088	0.4744	1886	0.4411	1628	0.5026	1763	0.4586
3132	0.4772	2829	0.4455	2442	0.5275	2645	0.4613
4176	0.4797	3772	0.4744	3256	0.5404	3527	0.4631
5220	0.4818	4715	0.4800	4070	0.5440	4408	0.4645
6264	0.4831	5658	0.4838	4884	0.5482	5290	0.4657
7308	0.4838	6601	0.4862	5698	0.5519	6172	0.4666
8352	0.4852	7544	0.4882	6512	0.5550	7053	0.4676
9396	0.4857	8487	0.4899	7326	0.5584	7935	0.4681
10,440	0.4869	9430	0.4913	8140	0.5613	8817	0.4689
11,484	0.4876	10,373	0.4927	8954	0.5639	9698	0.4693
12,528	0.4884	11,316	0.4937	9768	0.5664	10,580	0.4697
13,572	0.4890	12,259	0.4948	10,582	0.5693	11,462	0.4704
14,616	0.4897	13,202	0.4958	11,396	0.5713	12,345	0.4708
15,660	0.4902	14,145	0.4927	12,210	0.5741	13,225	0.4714
16,704	0.4906	15,088	0.4940	13,024	0.5764	14,107	0.4715
17,748	0.4910	16,031	0.4948	13,838	0.5787	14,989	0.4720
18,792	0.4913	16,974	0.4961	14,652	0.5813	15,871	0.4725
19,836	0.4919	17,917	0.5094	15,466	0.5842	16,752	0.4733
RS-5-3 (1.0 MPa, 5 Hz)		RS-6-2 (2.5 MPa, 5 Hz)		RS-7-1 (5.0 MPa, 5 Hz)		RS-8-3 (1.0 MPa, 10 Hz)	
Number of disturbance cycles	Maximum axial strain	Number of disturbance cycles	Maximum axial strain	Number of disturbance cycles	Maximum axial strain	Number of disturbance cycles	Maximum axial strain
1	0.4449	1	0.3802	1	0.3352	1	0.4249
759	0.4542	199	0.3900	623	0.4762	53	0.4399
1518	0.4595	419	0.3915	1246	0.4818	106	0.4425
2277	0.4631	628	0.3929	1869	0.4854	159	0.4442
3036	0.4662	838	0.394	2492	0.4881	212	0.4455
3795	0.4683	1047	0.3949	3115	0.4904	265	0.4479
4554	0.4704	1257	0.3974	3738	0.4922	318	0.4514
5313	0.4723	1466	0.4003	4361	0.4939	371	0.4550
6072	0.4740	1676	0.4035	4984	0.4952	424	0.4589

(continued)

Table 4.5 (continued)

RS-5-3 (1.0 MPa, 5 Hz)		RS-6-2 (2.5 MPa, 5 Hz)		RS-7-1 (5.0 MPa, 5 Hz)		RS-8-3 (1.0 MPa, 10 Hz)	
Number of disturbance cycles	Maximum axial strain	Number of disturbance cycles	Maximum axial strain	Number of disturbance cycles	Maximum axial strain	Number of disturbance cycles	Maximum axial strain
6831	0.4755	1885	0.4068	5607	0.4966	477	0.4632
7590	0.4769	2095	0.4103	6230	0.4978	530	0.4674
8349	0.4785	2304	0.4139	6853	0.4988	583	0.4699
9108	0.4796	2514	0.4178	7476	0.4999	636	0.4716
9867	0.4806	2723	0.4220	8099	0.5009	689	0.5055
10,626	0.4821	2933	0.4262	8722	0.5017	742	0.5184
11,385	0.4834	3142	0.4305	9345	0.5024	795	0.5239
12,144	0.4844	3352	0.4351	9968	0.5031	848	0.5304
12,903	0.4858	3561	0.4380	10,591	0.5038	901	0.5370
13,662	0.4868	3771	0.4400	11,214	0.5050	954	0.5418
14,421	0.4895	3980	0.4422	11,837	0.5056	1007	0.5468

variable based on maximum volumetric strain—this approach was not adopted. The cyclic dynamic disturbance tests were conducted at a stress level of 85% of the peak strength, a condition where the rock specimens were either at the terminal stage of linear elastic behavior or the initial stage of nonlinear deformation. At this stress level, the rock can no longer be reasonably treated as a purely elastic material. Consequently, the volumetric strain calculated using elastic mechanics theory would lack sufficient accuracy, and therefore, the damage variable based on maximum volumetric strain was excluded from further discussion.

During the cyclic dynamic disturbance process for each typical rock specimen in this experiment, 20 data points were uniformly selected, with statistical analysis conducted to determine the corresponding disturbance cycles and their associated maximum axial strains (presented in Table 4.5) and maximum radial strains (documented in Table 4.6) at each sampling point. This systematic data collection approach ensures comprehensive characterization of strain evolution throughout the loading history while maintaining consistent spatial and temporal resolution across all test specimens.

As illustrated in Figs. 4.12 and 4.13, B-spline approximation was applied to the selected dataset to characterize the evolution of both maximum axial and radial strains with increasing cycle numbers under various orthogonal combinations of three amplitude levels and three frequency levels (excluding the specific case of 20 Hz frequency with 5.0 MPa amplitude). The results demonstrate that both maximum axial and radial strains exhibit similar evolutionary patterns: they increase progressively with continued cyclic micro-disturbance loading, while showing characteristically rapid growth during initial loading stages followed by gradually decreasing growth rates across all tested amplitude and frequency conditions. This consistent

Table 4.6 Number of disturbance cycles and the maximum radial strain (incomplete statistics)

RS-1-1 (1.0 MPa, 5 Hz)		RS-2-2 (2.5 MPa, 5 Hz)		RS-3-1 (5.0 MPa, 5 Hz)		RS-4-3 (1.0 MPa, 10 Hz)	
Number of disturbance cycles	Maximum radial strain	Number of disturbance cycles	Maximum radial strain	Number of disturbance cycles	Maximum radial strain	Number of disturbance cycles	Maximum radial strain
1	0.06245	1	0.05901	1	0.04953	1	0.03482
1044	0.06322	943	0.06589	814	0.05510	882	0.06010
2088	0.06774	1886	0.06863	1628	0.07614	1763	0.06436
3132	0.07041	2829	0.07060	2442	0.10485	2645	0.06653
4176	0.07296	3772	0.08346	3256	0.11459	3527	0.06812
5220	0.07620	4715	0.08945	4070	0.12159	4408	0.06933
6264	0.07901	5658	0.09307	4884	0.12745	5290	0.07041
7308	0.08015	6601	0.09581	5698	0.13261	6172	0.07130
8352	0.08123	7544	0.09817	6512	0.13732	7053	0.07207
9396	0.08162	8487	0.10014	7326	0.14152	7935	0.07277
10,440	0.08378	9430	0.10186	8140	0.14534	8817	0.07340
11,484	0.08384	10,373	0.10339	8954	0.14916	9698	0.07404
12,528	0.08442	11,316	0.10485	9768	0.15304	10,580	0.07474
13,572	0.08505	12,259	0.10612	10,582	0.15712	11,462	0.07550
14,616	0.08575	13,202	0.10733	11,396	0.16119	12,345	0.07589
15,660	0.08633	14,145	0.10364	12,210	0.16508	13,225	0.07633
16,704	0.08677	15,088	0.10504	13,024	0.16877	14,107	0.07671
17,748	0.08728	16,031	0.10638	13,838	0.17265	14,989	0.07678
18,792	0.08766	16,974	0.10759	14,652	0.17653	15,871	0.07709
19,836	0.08843	17,917	0.10886	15,466	0.18061	16,752	0.07748
RS-5-3 (1.0 MPa, 5 Hz)		RS-6-2 (2.5 MPa, 5 Hz)		RS-7-1 (5.0 MPa, 5 Hz)		RS-8-3 (1.0 MPa, 10 Hz)	
Number of disturbance cycles	Maximum radial strain	Number of disturbance cycles	Maximum radial strain	Number of disturbance cycles	Maximum radial strain	Number of disturbance cycles	Maximum radial strain
1	0.07754	1	0.03126	1	0.02890	1	0.06042
759	0.09505	199	0.03772	623	0.07881	53	0.06697
1518	0.10199	419	0.03887	1246	0.08970	106	0.06926
2277	0.10663	628	0.03963	1869	0.09626	159	0.07092
3036	0.11039	838	0.04033	2492	0.10046	212	0.07219
3795	0.11357	1047	0.04084	3115	0.10371	265	0.07347
4554	0.11644	1257	0.04154	3738	0.10632	318	0.07518
5313	0.11898	1466	0.04243	4361	0.10867	371	0.08282
6072	0.12128	1676	0.04351	4984	0.11071	424	0.08327

(continued)

Table 4.6 (continued)

RS-5-3 (1.0 MPa, 5 Hz)		RS-6-2 (2.5 MPa, 5 Hz)		RS-7-1 (5.0 MPa, 5 Hz)		RS-8-3 (1.0 MPa, 10 Hz)	
Number of disturbance cycles	Maximum radial strain	Number of disturbance cycles	Maximum radial strain	Number of disturbance cycles	Maximum radial strain	Number of disturbance cycles	Maximum radial strain
6831	0.12338	1885	0.04472	5607	0.11249	477	0.08473
7590	0.12541	2095	0.04619	6230	0.11415	530	0.08703
8349	0.12739	2304	0.04771	6853	0.11561	583	0.08894
9108	0.12930	2514	0.04950	7476	0.11695	636	0.09046
9867	0.13121	2723	0.05141	8099	0.11822	689	0.11421
10,626	0.13312	2933	0.05364	8722	0.11937	742	0.13057
11,385	0.13490	3142	0.05599	9345	0.12051	795	0.14037
12,144	0.13668	3352	0.05847	9968	0.12166	848	0.14922
12,903	0.13846	3561	0.06083	10,591	0.12268	901	0.15744
13,662	0.14018	3771	0.06268	11,214	0.12370	954	0.16578
14,421	0.14356	3980	0.06420	11,837	0.12567	1007	0.17373

behavioral pattern suggests a fundamental response mechanism of rock specimens to cyclic micro-disturbance loading, independent of specific loading parameters.

Figure 4.14 presents the relationship curves between the modified damage variables (defined by both the maximum axial strain method and maximum radial strain method) and relative cycle number. The results demonstrate that these strain-based methods effectively characterize the damage degradation process of rock specimens under cyclic micro-disturbance loading, while appropriately accounting for the initial damage state at the commencement of loading. As evidenced by the curve trends, both damage variables exhibit fundamentally similar evolutionary patterns: (1) rapid initial growth with increasing relative cycle number during early loading stages, followed by (2) progressively decreasing growth rates in subsequent phases. This consistent behavior suggests that either strain component (axial or radial) can serve as a reliable indicator for quantifying damage accumulation in rocks subjected to cyclic micro-disturbances.

(D) Comprehensive Discussion

The comprehensive analysis of the three proposed methods for characterizing rock damage evolution under cyclic dynamic disturbance and high-stress conditions demonstrates that the selected physical parameters (residual strain, average dynamic axial stiffness, and maximum strain) effectively satisfy four fundamental requirements for damage variable definition: possessing clear physical significance, accounting for initial specimen damage, demonstrating evolutionary consistency with the damage degradation process, and maintaining theoretical robustness. These characteristics validate the scientific rationale behind employing these strain-based parameters to quantify damage progression during cyclic dynamic disturbance loading. However, several important considerations emerge when examining the practical implementation and limitations of these methods.

While all three parameters fundamentally rely on strain measurements derived from axial and radial deformation data, their interpretation for brittle and quasi-brittle rock materials requires careful consideration. The strain responses may potentially reflect mechanisms such as pore space compaction or crack closure rather than genuine damage to the rock matrix, particularly in high-porosity rock specimens. This limitation becomes especially relevant when attempting to extend the methodology beyond the specific experimental conditions of this study, where specimens were pre-loaded to approximately 85% of their uniaxial compressive strength before applying cyclic dynamic disturbances. Within these controlled laboratory conditions, the three parameters demonstrate satisfactory performance in characterizing damage evolution.

The methods differ significantly in their approaches to determining initial damage conditions. The residual strain method offers direct calculation through strain ratios at different loading stages, while both the dynamic stiffness and maximum strain methods require modification of their fundamental equations and depend on externally obtained initial damage values. In this study, these initial values were derived from a constitutive model developed in Sect. 6.2 based on quasi-static loading tests. Although this approach provides accurate results for the current experimental conditions, its broader application to different rock types or loading conditions would require re-calibration of model parameters through additional experimental data, adding complexity to the implementation process.

From a practical engineering perspective, implementing these damage characterization methods in field conditions presents considerable challenges. Real-world monitoring of rock deformation typically lacks the precision required to track continuous strain evolution under disturbance loading. The difficulty in obtaining real-time data on critical disturbance parameters (including frequency, waveform characteristics, and amplitude) under field conditions, combined with the necessity for laboratory verification of damage quantification, currently limits these methods primarily to laboratory research applications. While they provide valuable insights into damage mechanisms and establish important benchmarks for rock behavior under cyclic disturbances, their direct application to engineering practice requires significant advancements in field monitoring technologies and data acquisition systems. The methods nevertheless serve as an important foundation for developing more practical damage evaluation approaches suitable for engineering implementation.

4.3.4 Failure Modes After Disturbance Loading Tests

Figure 4.15a illustrates failure modes of rock samples after disturbance loading tests under coupling conditions of three different levels of frequencies and three different levels of amplitudes. According to these photographs, the failure modes of rock samples after disturbance loading tests can be sorted into four kinds as follows:

(a) Disturbance tests

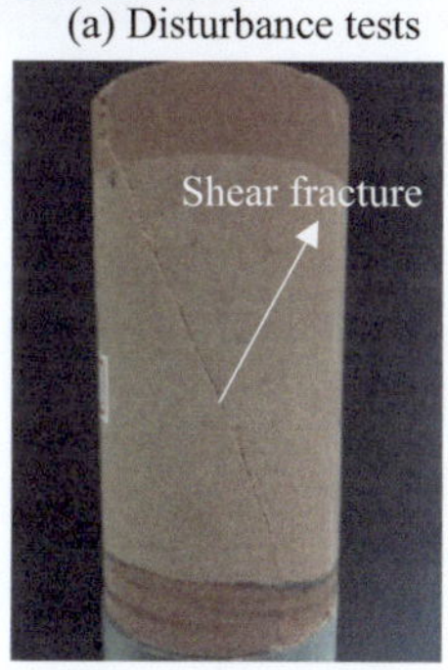

(b) Uniaxial compression test

Fig. 4.15 Failure modes of rock samples

1. In the condition of frequency of 20 Hz and amplitude of 5 MPa (shown in Fig. 4.15a-ix), the rock sample failed by a single inclined macroscopic shear fracture with no other obvious derivative cracks, which is approximately that of failure mode under uniaxial compression (shown in Fig. 4.15b). The failure modes attribute completely brittle. From Fig. 4.2 and Table 4.2, it is known that the rock sample in this case fails once the axial disturbance load is applied, and the number of cycles equals to zero. Essentially, the maximum stress on the rock sample with the disturbance amplitude of 5.0 MPa is very close to the peak stress of rock sample under uniaxial compression. In this amplitude condition, the disturbance frequency of 20 Hz, which is a relatively high level and corresponds to a relatively high strain rate, would cause sudden failure of the rock sample, which is similar to the process of impact failure of rock sample (Li et al. 2018).
2. Under coupling conditions of relatively lower frequencies and relatively lower amplitudes, including Fig. 4.15a-i, a-ii, a-iii, a-iv, and a-vii, the rock samples fail by an inclined macroscopic shear fracture with a few numbers of derivative cracks around it. With the increase of frequency and amplitude, the surfaces of rock samples appear to be more flaked and exfoliated at local positions. The failure modes still attribute brittle to some extent.
3. In the case of frequency of 10 Hz and amplitude of 2.5 MPa (shown in Fig. 4.15a-v), no obvious macroscopic shear fracture can be seen in the damaged samples, and a great number of derivative cracks and flaked and exfoliated pieces are spread all over on the surface of the rock sample. The fragmentation level is relatively higher than the abovementioned two cases. The failure of rock sample exhibits semibrittle (Evans et al. 1990).
4. When the disturbance load is applied with the frequency of 10 Hz coupling with the amplitude of 5.0 MPa, the rock specimen fails by the mode of cataclasis and exhibits brittle-ductile transition.

In summary, the failure modes of rock samples under relatively lower frequencies and amplitudes coupling conditions characterize shear failure with a main inclined fracture and a few numbers of derivative cracks around; with the coupling frequency and amplitude increase, the fragmentation degree grows, the brittleness of failure weakens and the ductility strengthens; the failure modes under coupling conditions of relatively higher frequencies and amplitudes attribute cataclasis and exhibit some ductile characteristics.

4.4 Chapter Summary

This chapter investigated the progressive damage and failure of rock under both external cyclic disturbance load and internal stress using the newly modified rock triaxial test system described in Chap. 2. A series of disturbance loading tests with different disturbance frequencies (5 Hz, 10 Hz, and 20 Hz) and different disturbance amplitudes (1.0 MPa, 2.5 MPa, and 5.0 MPa) was carried out on sandstone samples at the axial stress level of 85% of uniaxial compressive strength. The following conclusions are drawn from the experimental results:

1. The frequency and amplitude of the axial disturbance load have important effects on time duration of disturbance loading phase. Specifically, with increased amplitude at a same frequency, the time duration of disturbance loading decreases. When the disturbance loading is applied at the same level of amplitude, the time duration also decreases with the increased frequency.
2. The changing trends of axial and radial strain versus time are extremely similar with regime I and regime II—transient creep and steady-state creep—in idealized one-dimensional creep curve.
3. Damage variable defined by residual strain method, average dynamic axial stiffness method, and maximum axial and radial strain method can not only well reflect the gradual increase in damage degree of rock samples in disturbance loading tests under different disturbance frequencies and different amplitudes conditions, but can also characterize the initial damage degree of the disturbance loading phase.
4. The failure modes of rock samples under relatively lower frequencies and amplitudes coupling conditions characterize shear failure with a main inclined fracture and a few number of derivative cracks around; with the disturbance frequency and amplitude increase, the fragmentation degrees grow, the brittleness of failure weakens and the ductility strengthens; the failure modes under coupling conditions of relatively higher frequencies and amplitudes attribute cataclasis and exhibit some ductile characteristics.

References

Akai K, Ohnishi Y (1983) Strength and deformation characteristics of soft sedimentary rock under repeated and creep loading. In: 5th ISRM Congress

Areias P, Msekh MA, Rabczuk T (2016) Damage and fracture algorithm using the screened Poisson equation and local remeshing. Eng Fract Mech 158(6):116–143. https://doi.org/10.1016/j.engfracmech.2015.10.042

Areias P, Reinoso J, Camanho PP et al (2018) Effective 2D and 3D crack propagation with local mesh refinement and the screened Poisson equation. Eng Fract Mech 189(2):339–360. https://doi.org/10.1016/j.engfracmech.2017.11.017

Bek YK, Hamdia KM, Rabczuk T et al (2018) Micromechanical model for polymeric nanocomposites material based on SBFEM. Compos Struct 194(6):516–526. https://doi.org/10.1016/j.compstruct.2018.03.064

Brace WF, Paulding BW, Scholz C (1966) Dilatancy in the fracture of crystalline rocks. J Geophys Res 71(16):3939–3953. https://doi.org/10.1029/JZ071i016p03939

Cai M, Kaiser PK, Tasaka Y et al (2004) Generalized crack initiation and crack damage stress thresholds of brittle rock masses near underground excavations. Int J Rock Mech Min Sci 41(5):833–847. https://doi.org/10.1016/j.ijrmms.2004.02.001

Deng HF, Hu Y, Li JL et al (2017) Effects of frequency and amplitude of cyclic loading on the dynamic characteristics of sandstone. Rock Soil Mech 38(12):3402–3409 (in Chinese). https://doi.org/10.16285/j.rsm.2017.12.002

Evans B, Fredrich JT, Wong TF (1990) The brittle-ductile transition in rocks: recent experimental and theoretical progress. American Geophysical Union (AGU). https://doi.org/10.1029/GM056p0001

Fan JY, Chen J, Jiang DY et al (2016) Fatigue properties of rock salt subjected to interval cyclic pressure. Int J Fatigue 90:109–115. https://doi.org/10.1016/j.ijfatigue.2016.04.021

Gischig V, Preisig G, Eberhardt E (2016) Numerical investigation of seismically induced rock mass fatigue as a mechanism contributing to the progressive failure of deep-seated landslides. Rock Mech Rock Eng 49(6):2457–2478. https://doi.org/10.1007/s00603-015-0821-z

Jaeger JC, Cook NGW, Zimmerman R (2007) Fundamentals of rock mechanics, 4th edn. Wiley-Blackwell, Hoboken

Lemaitre J, Desmorat R (2005) Engineering damage mechanics: ductile, creep, fatigue and brittle failures. Springer, Berlin

Li XB, Gong FQ, Tao M et al (2017) Failure mechanism and coupled static-dynamic loading theory in deep hard rock mining: a review. J Rock Mech Geotech Eng 9(4):767–782. https://doi.org/10.1016/j.jrmge.2017.04.004

Li SH, Zhu WC, Niu LL et al (2018) Dynamic characteristics of green sandstone subjected to repetitive impact loading: phenomena and mechanisms. Rock Mech Rock Eng 51:1921–1936. https://doi.org/10.1007/s00603-018-1449-6

Liu EL, He SM (2012) Effects of cyclic dynamic loading on the mechanical properties of intact rock samples under confining pressure conditions. Eng Geol 125(27):81–91. https://doi.org/10.1016/j.enggeo.2011.11.007

Liu EL, He SM, Xue X et al (2011) Dynamic properties of intact rock samples subjected to cyclic loading under confining pressure conditions. Rock Mech Rock Eng 44(5):629–634. https://doi.org/10.1007/s00603-011-0151-8

Liu EL, Huang RQ, He SM (2012) Effects of frequency on the dynamic properties of intact rock samples subjected to cyclic loading under confining pressure conditions. Rock Mech Rock Eng 45(1):89–102. https://doi.org/10.1007/s00603-011-0185-y

Liu Y, Dai F, Zhao T et al (2017) Numerical investigation of the dynamic properties of intermittent jointed rock models subjected to cyclic uniaxial compression. Rock Mech Rock Eng 50(1):89–112. https://doi.org/10.1007/s00603-016-1085-y

Liu Y, Dai F, Dong L et al (2018) Experimental investigation on the fatigue mechanical properties of intermittently jointed rock models under cyclic uniaxial compression with different loading parameters. Rock Mech Rock Eng 51(1):47–68. https://doi.org/10.1007/s00603-017-1327-7

Ma L, Liu X, Wang M et al (2013) Experimental investigation of the mechanical properties of rock salt under triaxial cyclic loading. Int J Rock Mech Min Sci 62:34–41. https://doi.org/10.1016/j.ijrmms.2013.04.003

Mazaira A, Konicek P (2015) Intense rockburst impacts in deep underground construction and their prevention. Can Geotech J 52(10):1426–1439. https://doi.org/10.1139/cgj-2014-0359

Rabczuk T, Belytschko T (2004) Cracking particles: a simplified meshfree method for arbitrary evolving cracks. Int J Numer Meth Eng 61(13):2316–2343. https://doi.org/10.1002/nme.1151

Rabczuk T, Ren H (2017) A peridynamics formulation for quasi-static fracture and contact in rock. Eng Geol 225:42–48. https://doi.org/10.1016/j.enggeo.2017.05.001

Rabczuk T, Zi G, Bordas S et al (2010) A simple and robust three-dimensional cracking-particle method without enrichment. Comput Methods Appl Mech Eng 199(37):2437–2455. https://doi.org/10.1016/j.cma.2010.03.031

Ren S, Bai YM, Zhang JP et al (2013) Experimental investigation of the fatigue properties of salt rock. Int J Rock Mech Min Sci 64:68–72. https://doi.org/10.1016/j.ijrmms.2013.08.023

Ren H, Zhuang X, Cai Y et al (2016) Dual-horizon peridynamics. Int J Numer Meth Eng 108(12):1451–1476. https://doi.org/10.1002/nme.5257

Ren H, Zhuang X, Rabczuk T (2017) Dual-horizon peridynamics: a stable solution to varying horizons. Comput Methods Appl Mech Eng 318(1):762–782. https://doi.org/10.1016/j.cma.2016.12.031

Royer-Carfagni G, Salvatore W (2000) The characterization of marble by cyclic compression loading: experimental results. Int J Numer Anal Meth Geomech 5(7):535–563. https://doi.org/10.1002/1099-1484(200010)5:7%3c535::AID-CFM102%3e3.0.CO;2-D

Schijve J (2009) Fatigue of structures and materials, 2nd edn. Springer, Berlin

Taheri A, Royle A, Yang Z et al (2016) Study on variations of peak strength of a sandstone during cyclic loading. Geomech Geophys Geo-Energy Geo-Resour 2(1):1–10. https://doi.org/10.1007/s40948-015-0017-8

Tang CA (1997) Numerical simulation of progressive rock failure and associated seismicity. Int J Rock Mech Min Sci 34(2):249–261. https://doi.org/10.1016/S0148-9062(96)00039-3

Wang JA, Park HD (2002) Fluid permeability of sedimentary rocks in a complete stress–strain process. Eng Geol 63(3):291–300. https://doi.org/10.1016/S0013-7952(01)00088-6

Wang Z, Li S, Qiao L et al (2013) Fatigue behavior of granite subjected to cyclic loading under triaxial compression condition. Rock Mech Rock Eng 46(6):1603–1615. https://doi.org/10.1007/s00603-013-0387-6

Wang H, Liu D, Cui Z et al (2016) Investigation of the fracture modes of red sandstone using XFEM and acoustic emissions. Theor Appl Fract Mech 85(Part B):283–293. https://doi.org/10.1016/j.tafmec.2016.03.012

Xiao JQ, Ding DX, Jiang FL et al (2010) Fatigue damage variable and evolution of rock subjected to cyclic loading. Int J Rock Mech Min Sci 47:461–468. https://doi.org/10.1016/j.ijrmms.2009.11.003

Yang SQ, Jing HW, Wang SY (2012) Experimental investigation on the strength, deformability, failure behavior and acoustic emission locations of red sandstone under triaxial compression. Rock Mech Rock Eng 45(4):583–606. https://doi.org/10.1007/s00603-011-0208-8

Zhang Y, Lackner R, Zeiml M et al (2015) Strong discontinuity embedded approach with standard SOS formulation: Element formulation, energy-based crack-tracking strategy, and validations. Comput Methods Appl Mech Eng 287(4):335–366. https://doi.org/10.1016/j.cma.2015.02.001

Zhou ZL, Li XB, Zou Y et al (2014) Dynamic Brazilian tests of granite under coupled static and dynamic loads. Rock Mech Rock Eng 47(2):495–505. https://doi.org/10.1007/S00603-013-0441-4

Zhou S, Zhu HH, Ju JW et al (2017) Modeling microcapsule-enabled self-healing cementitious composite materials using discrete element method. Int J Damage Mech 26(2):340–357. https://doi.org/10.1177/1056789516688835

Zhu HH, Zhou S, Yan ZG, Ju JW et al (2015a) A two-dimensional micromechanical damage-healing model on microcrack-induced damage for microcapsule-enabled self-healing cementitious composites under tensile loading. Int J Damage Mech 24(1):95–115. https://doi.org/10.1177/1056789514522503

Zhu HH, Zhou S, Yan ZG et al (2015b) A 3D analytical model for the probabilistic characteristics of self-healing model for concrete using spherical microcapsule. Comput Concr 15(1):37–54. https://doi.org/10.12989/cac.2015.15.1.037

Zhu HH, Zhou S, Yan ZG et al (2016) A two-dimensional micromechanical damage-healing model on microcrack induced damage for microcapsule-enabled self-healing cementitious composites under compressive loading. Int J Damage Mech 25(5):727–749. https://doi.org/10.1177/1056789516641593

Bagde MN, Petroš V (2005a) Fatigue properties of intact sandstone samples subjected to dynamic uniaxial cyclical loading. Int J Rock Mech Min Sci 42(2):237-250. https://doi.org/10.1016/j.ijrmms.2004.08.008

Bagde MN, Petroš V (2005b) Waveform effect on fatigue properties of intact sandstone in uniaxial cyclical loading. Rock Mech Rock Eng 38(3):169-196. https://doi.org/10.1007/s00603-005-0045-8

Chapter 5
Effect of Confining Pressure on Damage Evolution and Failure Behaviors of Intact Sandstone Samples During Cyclic Dynamic Disturbance

5.1 Introduction

In deeply buried rock engineering, mechanical drilling, remote blasting, continuous low-frequency vibration of heavy machinery, and natural earthquakes often cause cyclic disturbance of the surrounding rocks. These cyclic disturbance loads have the characteristics of long transmission distances and high carrying energies, which may trigger damage, failure, or instability of the surrounding rocks (Zhu et al. 2010; Du et al. 2016). Under these cyclic disturbances, the free faces of the surrounding rock that simultaneously bears a high geostress may pose potential engineering hazards related to large deformation or displacement (Kaiser et al. 2001), such as swelling, spalling, breaking, and relaxing of the free faces of the surrounding rock. Considering that there is still no universal agreement as to how high geostress affects the characteristics of rock deformation, damage, and failure during cyclic disturbance, it is hard to accurately predict and effectively prevent and control those potential disturbance-triggered engineering hazards, which may pose a risk to workers' safety and cause economic losses. In laboratory tests, rock mechanics researchers usually apply confining pressure and axial stress to simulate the in-situ three-dimensional geostress of the surrounding rocks and to explore the characteristics of rock deformation, damage, and failure (Takemura and Oda 2005). Therefore, determining the effect of confining pressure on the damage evolution and failure behaviors of intact red sandstone samples during cyclic disturbance would be greatly beneficial to improving the design and construction level and to reducing engineering disasters in deeply buried rock projects.

Since the pioneering triaxial tests conducted by von Kármán (1911) on marble samples, numerous experimental studies have been conducted on the ability of confining pressure to clarify the characteristics of rock deformation, damage, and failure. Paterson (1958) conducted triaxial compressive tests on samples of Wombeyan marble at various confining pressures of up to 1000 kg cm^{-2}. The maximum stress increased with increasing confining pressure from 0 to about

X. Chen, *Damage and Failure of Rock Triggered by Dynamic Disturbance*,
https://doi.org/10.1007/978-981-96-9486-0_5

300 kg cm^{-2}, and the stress–strain curves rose continuously without reaching a maximum at a confining pressure of greater than about 300 kg cm^{-2}. This phenomenon was defined as the brittle-ductile transition, and it has been supported by many other scholars such as Simpson (1985), Wong and Baud (2012), and Baud et al. (2009). Christensen and Wang (1985) concentrated their efforts on understanding the effect of confining pressure and pore pressure on the dynamic elastic properties of Berea sandstone. They reported that the compressional- and shear-wave velocities of water-saturated Berea sandstone are strongly influenced by the confining and pore pressures. Li et al. (1999) comprehensively investigated the influence of the strain rate and confining pressure on a granite using a series of triaxial compression tests and found that the Young's modulus and Poisson's ratio seem to increase slightly with increasing confining pressure. Bésuelle et al. (2000) observed two distinct types of failure by localization from their 60 experiments of Vosges sandstone samples in triaxial compression under a large range of confining pressures (0–60 MPa), and they concluded that confining pressure has a strong influence on the localized deformation behaviors. In studying the influence of structural anisotropy on progressive damage of porous rock, Gatelier et al. (2002) conducted some cyclic compression tests under different confining pressures and loading orientations, and they found that the influence of anisotropy is significantly reduced with increased confining pressure. In a recent publication, Liu et al. (2017) studied the effect of confining pressure unloading on the strength reduction of soft coal using two different experimental stress paths. Their results revealed that the cohesion of the traditionally and newly reconstituted samples diminished by approximately 44.77% and 29.66%, respectively, with confining pressure unloading. Liang et al. (2017) investigated the effect of the continuous loading and unloading of confining pressure on the mechanical properties of a rock in the residual phase. They observed that the bearing capacity of the rock in the residual phase is related to both the confining pressure and the loading path. These previous studies basically focused on the deformation behaviors, strength characteristics, and failure modes of rocks subjected to confining pressure, and they provided some acceptable mechanical responses for rock samples under confined conditions.

In addition, several relevant investigations of the effects of the coupling of confining pressure and cyclic disturbance on the mechanical behaviors of rocks have been widely reported in recent years. To explore the influence of waveform, frequency and amplitude that are basic parameters to describe cyclic disturbance on mechanical properties of rock, Badge and Petroš (2005a, b) imposed dynamic uniaxial cyclical loads on sandstone samples. They concluded that loading waveforms strongly influenced the damage accumulation, and dynamic fatigue strength and dynamic stiffness reduced with frequencies and amplitude. Sun et al. (2017) proposed a rock fatigue damage model under multi-level amplitude cyclic loading with confining pressure based on a group of tests conducted on sandstone samples. He and Yang (2018) performed a series of laboratory tests to explore the deformation characteristics of rocks subjected to active confining pressure and cyclic blasting. It was found that a circumference compressive stress was produced by the confining pressure, which reduced the number and size of the radial cracks and the tensile failure through explosive loading. By performing conventional creep experiments, Chen et al. (2018a, b) studied the influence of unloading and loading stress cycles on the creep behavior of Darley Dale sandstone, and showed that dynamic stress perturbations transiently increase the strain rate, which could induce catastrophic and failure of long-term constant stressed rock. Zhang et al. (2018) reported that during cyclic triaxial loading on weathered red mudstone, the dynamic modulus, damping ratio, critical cyclic stress ratio, and dynamic stress level were extremely susceptible to the confining pressure and loading cycles. Peng et al. (2019) applied triaxial cyclic loading and unloading with an increasing lower cyclic stress limit on sandstone samples under confining pressures of 5, 10, and 15 MPa, and they concluded that increasing the confining pressures reduced the development of irreversible radial strain, increased the irreversible axial strain, and inhibited the development and connection of cracks. These abovementioned investigations indicate that the characteristics of rock samples in the loading path of cyclic loading and confining pressure, which is more in line with actual engineering loads, is extremely different from those in the loading path of conventional triaxial tests.

Despite the contributions of previous studies, the mechanism of the damage and failure of rocks under both triaxial stress and cyclic disturbance remains unclear. Therefore, a series of tests were performed on intact red sandstone samples under five different levels of confining pressures ranging from 0 to 40 MPa using the newly developed servo-controlled rock triaxial test system reported in Chap. 2. The main goals of the study in this chapter are as follows:

1. Gaining experimental data for axial cyclic disturbance tests in triaxial stress states of rock samples under different confining pressures.
2. Investigating the effect of confining pressure on the mechanical responses, damage evolution, and failure behaviors of intact sandstone samples subjected to cyclic disturbance systematically.

5.2 Testing Scheme and Procedure

The study in this chapter is a continuation of the content in Chap. 4, so the rock samples used in this study are produced in the same sandstone block as those used in the Chap. 4. The detailed processing process of the rock samples has been described in detail in Chap. 3 and will not be repeated here. As shown in Fig. 3.4 and Table 3.2, the compressive strengths of the sandstone samples with confining pressures of 0, 10, 20, 30, and 40 MPa were obtained at an axial displacement control rate of 0.02 mm/min about 64.70 MPa, 109.17 MPa, 139.12 MPa, 168.99 MPa, and 199.25 MPa, respectively.

In this study, tests with a continuous axial cyclic loading under a constant confining pressure were carried out. To assess the effects of confining pressure on the mechanical properties, five different levels of confining pressure, i.e., 0, 10, 20, 30, and 40 MPa, were applied. The experimental conditions of each rock sample in this study are shown in Table 5.1. After the confining pressures were applied, a two-phase loading method (shown in Fig. 5.1a) was used, i.e., a quasi-static loading phase and a cyclic disturbance phase. In the first phase, the quasi-static loading was monotonically increased under axial displacement-controlled conditions with a loading rate of 0.02 mm/min, i.e., the same as in the previously conventional triaxial compression tests. When the desired target set point was reached, i.e., 85% of the triaxial compressive strength, at each confining pressure, the deviatoric stress was maintained at this level. The target set point was chosen to be large enough to make it correspond to a relatively high level in the linear elastic stage on the stress–strain curve. In the second phase, by servo-controlling the disturbance loading rod so that it set on the top platen, cyclic disturbance was applied to the rock specimen in the axial direction. The periodic disturbance in this study was specified as a sinusoidal form with a frequency of 10 Hz and an amplitude of 2.5 MPa. The stress conditions during the cyclic disturbance phase are shown in Fig. 5.1b. This phase ended when the rock sample was cyclically loaded to failure.

Table 5.1 Experimental conditions of each rock sample

Sample no.	Confining pressure/MPa	Sample no.	Confining pressure/MPa
RS-10-1	10	RS-12-1	30
RS-10-2	10	RS-12-2	30
RS-10-3	10	RS-12-3	30
RS-11-1	20	RS-12-1	40
RS-11-2	20	RS-12-2	40
RS-11-3	20	RS-12-3	40

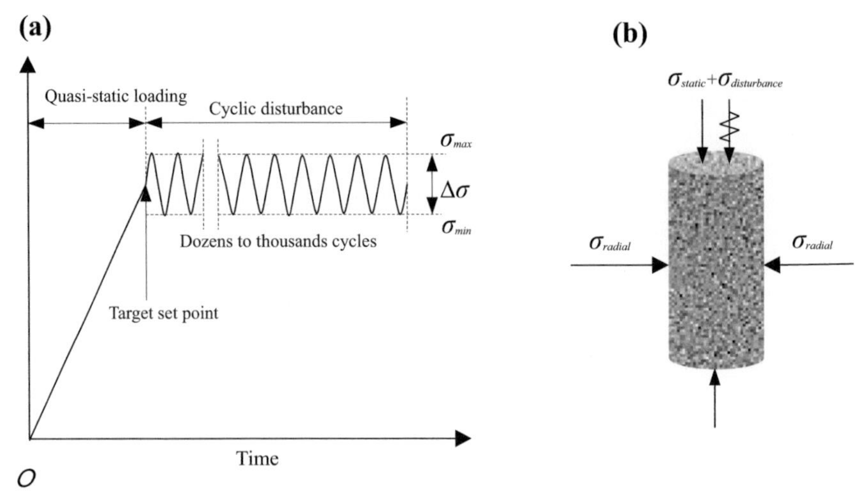

Fig. 5.1 Diagrammatic sketch of the loading method: **a** loading path and **b** stress conditions

To apply the confining pressure, the rock sample needs to be carefully sealed before loading because whether effective jacket sealing is achieved or not directly determines the success or failure of the tests. The rock sample was placed between the top and bottom platens of the testing machine to ensure that the central axis of the rock sample was aligned with the axis of the two platens. Then, the two platens and the rock sample were wrapped together with impermeable tape, which was properly stretched to enhance its stickiness. After this, an impervious Teflon heat-shrinkable tube, the diameter of which was slightly larger than that of the rock sample, was placed around the cylinder. The outer surface of the heat-shrinkable tube is evenly heated using a hot air blower to make the jacket closely adhere to the rock sample. The two ends of the heat-shrinkable tube were furtherly sealed with impermeable tape. As a result, the Teflon jacket prevented the confining fluid from entering the pores and cracks inside the rock sample.

After the rock sample was completely sealed, a circumferential and two axial extensometers (LVDT which manufactured by Jilin Province Jinli Test technology Co., Ltd.), with a measurement range of 4 mm and an accuracy of $\pm$ 0.001 mm, were installed, which allowed for the acquisitions of the axial strain $\boldsymbol{\varepsilon}_1$ and radial strain $\boldsymbol{\varepsilon}_3$. Each extensometer was used in conjunction with a one-fourth Wheatstone bridge. Subsequently, the assembled sample was placed on the loading platform of the testing machine. As is shown in Fig. 5.2, after the bolts on the flange of the bottom platen were tightened and the pipeline and data lines were connected, the assembled rock sample was used for the loading test. The thermal effects were also taken into consideration in this study either.

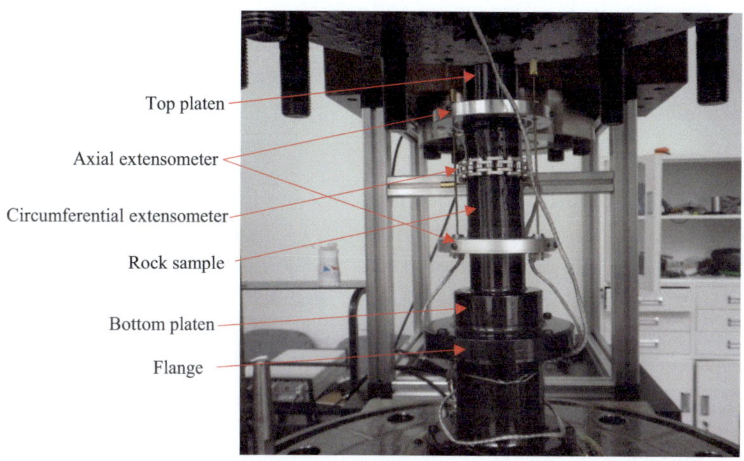

Fig. 5.2 Assembled rock sample

5.3 Experimental Results and Discuss

5.3.1 Characteristics of Stress–Time Curves and Stress–Strain Curves

Using the loading path shown in Fig. 5.1a, typical curves of the variation in the deviatoric stress with time (Fig. 5.3) were obtained from cyclic disturbance tests under five different confining pressures (i.e., 0, 10, 20, 30, and 40 MPa). As can be observed from Fig. 5.3a–e, the deviatoric stress-time curves of the quasi-static loading phase exhibit a similar variation tendency under all of the confining pressure conditions; that is, the deviatoric stress is initially slightly concave upward with increasing curvature as time increases, and then, it changes to a linear stage with a constant curvature, which correspond to the compaction stage and linear elastic stage, respectively. In the compaction stage, the primary micropores, microfissures, and flaws in the rock sample are compacted and closed under the relatively small three-dimensional stresses. In the linear elastic stage, the primary micropores and microfissures begin to propagate, and secondary cracks gradually form, grow, and steadily connect with increasing axial stress. This is evident from direct SEM observations (Wong 1982) and numerical simulation (e.g. RFPA and PFC) results (Kaiser and Tang 1998; Schöpfer et al. 2009).

Following the loading path, when the deviatoric stress-time curves enter the cyclic disturbance phase, as can also be seen from Fig. 5.3a–e, the length of time that the rock sample can bear the cyclic disturbance is quite susceptible to the confining pressure. Specifically, the time period and number of disturbance cycles before the rock sample fails increase with increasing confining pressure. At the end of the cyclic disturbance phase, the deviatoric stress exhibits a steep instantaneous decline, indicating that the cyclic disturbance triggers continuous accumulation of damage inside the rock sample, which eventually causes the rock sample to undergo brittle failure.

Corresponding to Figs. 5.3, 5.4 illustrates the variation in the deviatoric stress with axial strain and radial strain over the course of the loading of sandstone samples at different confining pressures of 0–40 MPa. In the quasi-static loading phase, the loading path and loading rate are the same as those in the monotonous triaxial compression tests, leading to the same curve variation tendencies shown in Fig. 3.4. In the disturbance loading phase, although a regular sinusoidal cyclic disturbance with a frequency of 10 Hz and an amplitude of 2.5 MPa was imposed on each sandstone sample in the axial direction, the obtained deviatoric stresses at the different confining pressures did not vary in the same imposed sinusoidal pattern as the axial and radial strains. During loading, the growths of these strains are mainly owing to formation, propagation, connection and nucleation of microcracks along weak grain boundaries (Cai et al. 2004). The deformation of the matrix of rock is usually considered to be incompressible or negligible. Thus, the irregular stress–strain responses indicate that the mechanical properties during cyclic disturbance are heavy reliant on the microscopic flaws within the rock that, making the rock materials exhibit

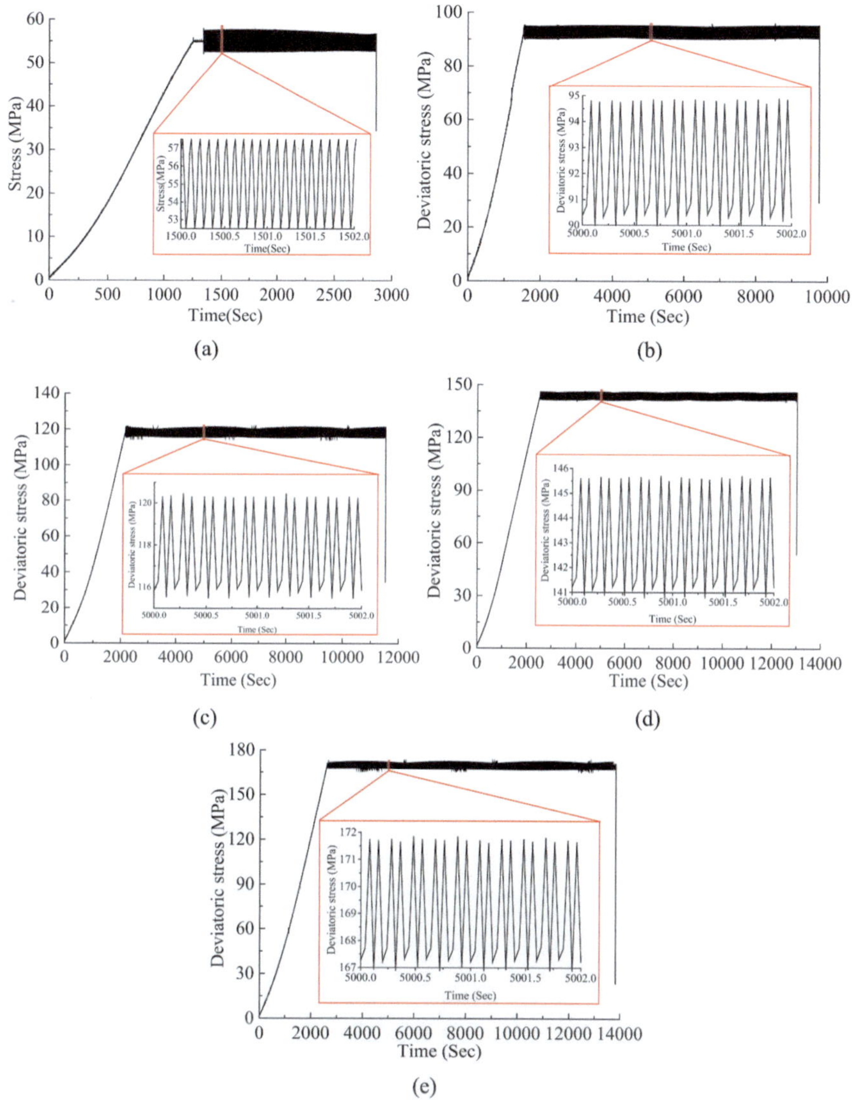

Fig. 5.3 Deviatoric stress–time curves obtained from the cyclic disturbance tests: **a** $\sigma_3 = 0$ MPa (RS-5-3), **b** $\sigma_3 = 10$ MPa (RS-10-1), **c** $\sigma_3 = 20$ MPa (RS-11-1), **d** $\sigma_3 = 30$ MPa (RS-12-2), **e** $\sigma_3 = 40$ MPa (RS-13-3)

characteristics of nonlinearity and inhomogeneity and differ from metal materials (Schijve 2009).

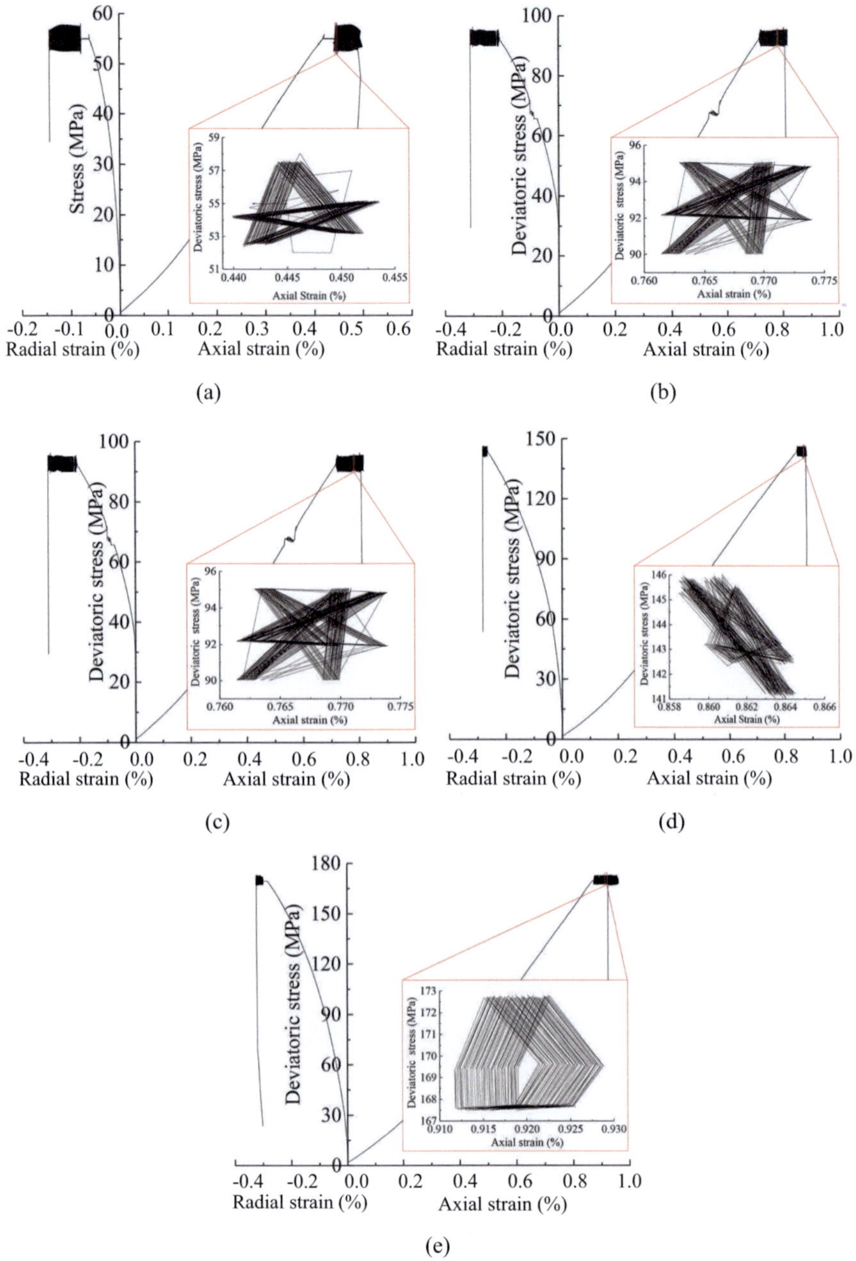

Fig. 5.4 Stress–strain curves obtained from the cyclic disturbance tests under different confining pressures: **a** $\sigma_3 = 0$ MPa (RS-5-3), **b** $\sigma_3 = 10$ MPa (RS-10-1), **c** $\sigma_3 = 20$ MPa (RS-11-1), **d** $\sigma_3 = 30$ MPa (RS-12-2), **e** $\sigma_3 = 40$ MPa (RS-13-3)

5.3.2 *Effect of Confining Pressure on the Mechanical Responses of the Sandstone Samples*

Table 5.2 shows the data obtained from the stress-time or stress–strain curves, including the duration of the dynamic disturbance, the number of disturbance cycles, the maximum axial and radial strains before failure of the sandstone samples, and the volumetric strains at the time of failure. As can be seen from the table, as the confining pressure increased, the duration of the cyclic disturbance increased, and the number of corresponding dynamic disturbance periods also increased. Figure 5.5a depicts the duration of the cyclic disturbance under the five different confining pressures. The duration does not change linearly with increasing confining pressure, but rather, it is concave toward the confining pressure axis, with a faster growth rate at relatively low confining pressures and a slower growth rate at relatively high confining pressures. The relationship between the number of disturbance cycles and the confining pressure is approximately exponential, and the specific functional form of this relationship can be fitting with

$$\begin{cases} N = 108939.935 - 93194.479e^{-0.11235\sigma_c} \\ R^2 = 0.97856 \end{cases} \tag{5.1}$$

Table 5.2 Data for the mechanical responses of the rock samples in the cyclic disturbance tests

Sample no.	Confining pressure (MPa)	Duration of cyclic disturbance (s)	Number of disturbance cycles	Maximum axial strain before failure (%)	Maximum radial strain before failure (%)	Volumetric strain at failure (%)
RS-5-1	0	1867	18,670	0.5583	– 0.1346	0.2730
RS-5-2	0	1382	13,820	0.4890	– 0.1259	0.2213
RS-5-3	0	1518	15,180	0.5001	– 0.1442	0.2037
RS-10-1	10	8228	82,280	0.8168	– 0.3102	0.1868
RS-10-2	10	7784	77,840	0.7903	– 0.2824	0.2157
RS-10-3	10	8361	83,610	0.8211	– 0.2657	0.2794
RS-11-1	20	9315	93,150	0.8450	– 0.2305	0.3779
RS-11-2	20	8650	86,500	0.8482	– 0.2780	0.2826
RS-11-3	20	9927	99,270	0.8408	– 0.2274	0.3708
RS-12-1	30	11,529	115,290	0.8830	– 0.2901	0.2924
RS-12-2	30	10,478	104,780	0.8794	– 0.2815	0.3121
RS-12-3	30	9998	99,980	0.8664	– 0.2882	0.2801
RS-13-1	40	11,350	113,500	1.0200	– 0.3045	0.4013
RS-13-2	40	10,612	106,120	0.9955	– 0.3336	0.3189
RS-13-3	40	11,172	111,720	0.9613	– 0.3237	0.2769

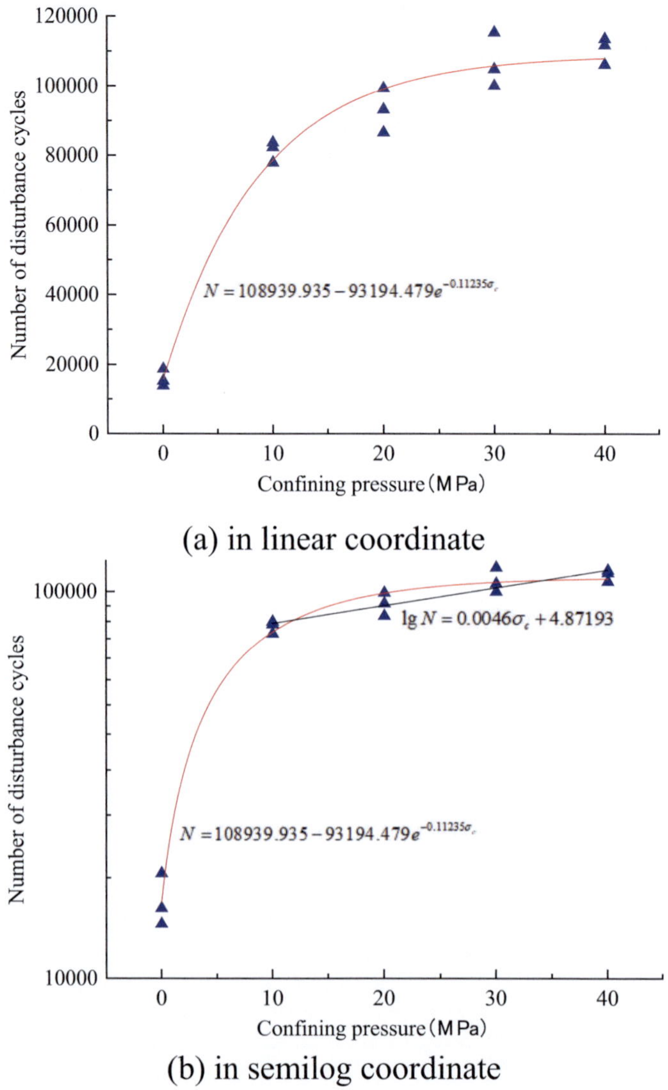

Fig. 5.5 Relationship between the number of disturbance cycles and the confining pressure

Figure 5.5a can be transformed into Fig. 5.5b with logarithmic y axis. It can be found that, except for the uniaxial compression, the logarithmic number of disturbance cycles against confining pressure is almost a horizontal line that has a linear fitting equation:

$$\lg N = 0.0046\sigma_c + 4.87193 \tag{5.2}$$

The reason for this phenomenon may be that when the rock sample is in unconfined condition, the brittleness makes the rock deform with less cumulative irreversible strain when cyclic disturbance is applied. When the confining pressure is applied, the ductility of the samples fosters cumulative irreversible strain. This cumulative irreversible strain does not increase linearly, but gradually stabilizes as the confining pressure increases. On the other hands, total-life of high-cycle fatigue or disturbance are generally presented in the form of an Wöhler's S-N (stress versus log number of cycles to failure) curve (Campbell 2012) which is given by

$$\lg N = a\Delta\sigma + b \tag{5.3}$$

where a and b are parameters related to material, stress ratio and loading method, and $\Delta\sigma$ is cyclic stress range as show in Fig. 5.1a. From the comparison, we know that Eqs. (5.2) and (5.3) have a similar form except the difference between $\boldsymbol{\sigma}_c$ and $\Delta\boldsymbol{\sigma}$. In this study, $\Delta\sigma$ is preset to reach and maintain at a constant value of 5.0 MPa, thus it seems to obtain same cycles to failure N under different levels of confining pressures. However, Fig. 3.4 shows that triaxial compressive strength $\boldsymbol{\sigma}_{TCS}$ becomes higher at higher confining pressure, indicating the ratio of triaxial compressive strength to cyclic stress range $\boldsymbol{\sigma}_{TCS}/\Delta\sigma$ increases with increasing confining pressure. Therefore, it can be boldly inferred that, under confining pressure conditions, the log number of disturbance cycles is roughly linear with the ratio of $\boldsymbol{\sigma}_{TCS}/\Delta\sigma$ and asymptotically approaches to a horizontal line with increasing confining pressure.

Figure 5.6 compares the maximum axial strain before the failure of each sandstone sample in the cyclic disturbance tests (blue bars) to the axial strains corresponding to the triaxial compressive strengths (TCS) (red bars) in the conventional triaxial tests, which are 0.526%, 0.950%, 1.022%, 1.230%, and 1.349% under confining pressures of 0, 10, 20, 30, and 40 MPa, respectively. As can be seen from this bar chart, under uniaxial compressive conditions (i.e., a confining pressure of 0 MPa), the axial strain corresponding to the TCS is close to the maximum axial strain, with a relative error of less than 10%. This indicates that the rock samples fail at similar axial strains under zero confining pressure and monotonic loading and cyclic disturbance, which makes the maximum normal strain criterion (i.e., Saint-Venant's criterion) an applicable criterion for judging whether the sandstone samples fail or not under unconfined conditions in the cyclic disturbance tests. In contrast, as is also shown by the bar chart, when the confining pressure of the sandstone samples is greater than zero, the maximum axial strains before failure under cyclic disturbance loading are less than the axial strains corresponding to the TCS under conventional loading conditions, and the differences between them (about 0.00141, 0.00177, 0.00354, 0.00356 in average) gradually increase with increasing confining pressure of 10, 20, 30 and 40 MPa, resulting in the inapplicability of the maximum normal strain criterion (i.e., Saint-Venant's criterion) to the cyclic disturbance tests with confining conditions. This is because Saint-Venant's criterion is especially suitable for brittle materials that fail suddenly through rupture or fracturing without any prior yielding when subjected to tensile or compressive stress (Beer et al. 2012), but the confining pressure weakens

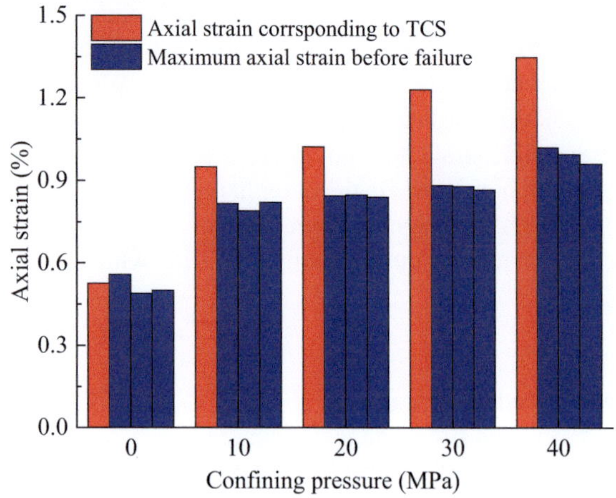

Fig. 5.6 Comparison of the axial strains corresponding to the triaxial compressive strength in conventional triaxial tests and the maximum axial strains before failure in the cyclic disturbance tests

the brittleness and enhances the ductility of the rock sample (Paterson and Wong 2005).

As can be seen from the last column in Table 5.2, the calculated volume strains ($\varepsilon_v = \varepsilon_a + 2\varepsilon_r$, where ε_v, ε_a and ε_r are respectively volume strain, axial strain and radial strain) at the failure of the rock samples are positive for all of the confining pressures tested; that is, before failure occurred in the cyclic disturbance tests, the rock samples did not undergo volume dilatation but break down during volume compression, which is different from the phenomenon observed in conventional triaxial compression tests (Brace et al. 1966). This indicates that the occurrences of the failure of the rock samples in the cyclic disturbance tests were sudden, which is also confirmed by Figs. 5.3 and 5.4, in which there are no obvious precursors to the failure. The suddenness of the failure of the rock samples under cyclic disturbance will be discussed further in Sect. 5.3.4.

5.3.3 *Effect of Confining Pressure on Damage Evolution*

According to the macroscopic phenomenological damage mechanics theory, the damage to solid materials is defined in order to study its discontinuous characteristics, including the initiation, growth of micropores and microfissures inside the material, and the mechanical behaviors of the damaged materials are described using methods for continuous mediums rather than for discontinuous mediums on the macroscopic scale (Rabczuk et al. 2010; Zhu et al. 2016). The damage evolution of solid materials can give rise to changes in the material's physical properties and the deterioration of its mechanical properties (Ougier-Simonin et al. 2011; Cao and Chung 2002; Oda et al. 1984), such as its porosity, crack density, ultrasonic wave velocity, acoustic emission cumulative counts, electrical resistivity, deformation modulus, and strength.

The pioneering definition of the damage variable was physically given by the surface density of the microcracks and the intersections of the microvoids on a plane cutting the representative volume element (RVE) of the cross section ΔS (Kachanov and Krajcinovic 1986), which is written as

$$D = \frac{\delta S_D}{\delta S} \tag{5.4}$$

where ΔS_D is the cross section of the microcracks and microvoids.

The damage variable can also be defined using the ratios of other physical or mechanical parameters (i.e., the ratio of the parameter after the damage to that before the damage). As for which parameter should be chosen as a good representation of the degree of damage to the rock materials, Xiao et al. (2010) believed that the damage variable defined should be able to reflect the processes of microcrack initiation, propagation, connection, and nucleation in rock materials under loading conditions, and they proposed four basic rules for selecting a parameter to define the damage variable. (1) The selected parameter needs to have a clear physical or mechanical significance; (2) the selected parameter needs to be convenient to measure and be easily applied to practical engineering; (3) the variation tendency of the selected parameter needs to be in good agreement with the actual damage and degradation processes of the rock materials; and (4) the damage variable defined needs to take the initial damage degree into consideration. Following these four rules, they proposed a method for defining the damage variable using the maximum strain:

$$D = \frac{\varepsilon_{\max}^{n} - \varepsilon_{\max}^{0}}{\varepsilon_{\max}^{f} - \varepsilon_{\max}^{0}} \tag{5.5}$$

where $\varepsilon_{\max}^{0}$, $\varepsilon_{\max}^{n}$, and $\varepsilon_{\max}^{f}$ are the initial maximum strain, the instantaneous maximum strain after n cycles, and the ultimate maximum strain, respectively. In this study, since the monotonous loading before the cyclic disturbance loading caused a certain extent of damage to the rock samples, Eq. (5.5) can be furtherly modified to take the initial degree of damage of the cyclic disturbance into consideration:

$$D_{\text{mod}} = D_0 + \frac{\varepsilon_{\max}^{n} - \varepsilon_{\max}^{0}}{\varepsilon_{\max}^{f} - \varepsilon_{\max}^{0}} \cdot (1 - D_0) \tag{5.6}$$

where D_{mod} is the modified damage variable; and D_0 is the initial damage variable at the onset of the cyclic disturbance phase.

The existence of confining pressure can change both the stress state and the mechanical behaviors of the rock sample. As was reported by Paterson and Wong (2005), Liu et al. (2017), and Wong et al. (2013), the confining pressure causes the rock materials to undergo a medium transition, from discontinuity to continuity, from brittleness to ductility, and from elasticity to viscosity. The changes in the stress state and mechanical properties will inevitably lead to changes in the evolution process of the rock damage. In order to explore the influence of confining pressure on the damage variable in the cyclic disturbance phase, Fig. 5.7 shows the function of the modified damage variable D_{mod} for the relative number of cycles n/N based on the maximum axial and radial strain methods under five different confining pressures, where N is the number of ultimate disturbance cycles.

As can be seen from Fig. 5.7a, which was plotted using the maximum axial strain method, the initial damage variable (the first value on each curve) gradually decreases with increasing confining pressure. This demonstrates that the confining pressure plays the role of impeding the formation and growth of cracks in the rock

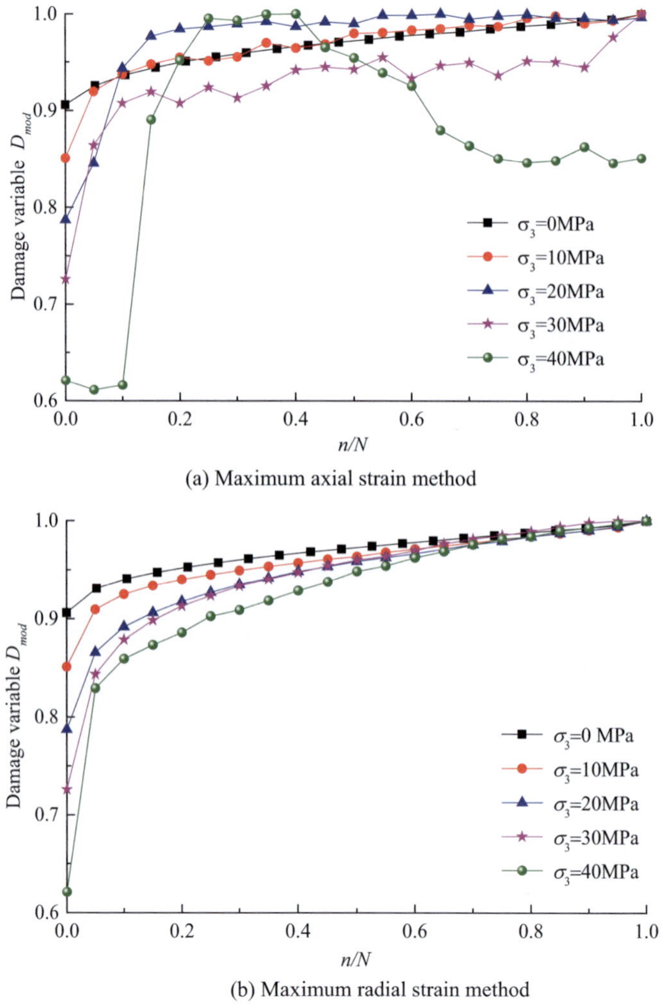

Fig. 5.7 Relationship between the modified damage variable D_{mod} and the relative cycles n/N

samples and reducing the degree of damage. The greater the confining pressure, the stronger the inhibition. Thus, it can be concluded that after the quasi-static loading phase, for the same loading rate, the rock samples are less damaged under greater confining pressures. Furthermore, during cyclic disturbance loading, the damage variables of the rock samples increase with a similar trend. That is, they initially grow dramatically with a steep slope, and then, they grow slowly with a gentle slope, indicating that the damage to the rock sample develops rapidly at the beginning of the cyclic disturbance phase but relatively slowly at the end of the phase. This is consistent with the phenomenon reported by Liu and He (2012), that is, as the number of loading cycles increased, the samples became elastic–plastic and developed irreversible deformation including axial, volumetric, and lateral strains, and the deformation magnitudes became greater. Moreover, there is no distinct rule describing the variation in the $D_{\text{mod}} - n/N$ curves with confining pressure based on the maximum axial strain method, whereas the overall level of the $D_{\text{mod}} - n/N$ curves plotted based on the maximum radial strain method becomes higher at higher confining pressures. The reason for this difference is that the test dispersion and the error in the axial strain measurement are greater than that of the radial strain measurement during cyclic disturbance.

5.3.4 Effect of Confining Pressure on Failure Results

Figure 5.8a, b show the failure results of the sandstone samples at the macroscopic scale after the conventional triaxial compression tests and the cyclic disturbance tests under the five different levels of confining pressures from 0 to 40 MPa, respectively. Based on the failure results after the conventional triaxial compression tests, all the sandstone samples failed through a single inclined macroscopic shear fracture. As the confining pressure increased, the angle between the fracture surface and the axial direction of the rock sample increased slightly. Chen et al. (2017) found that as the confining pressure increased, the failure of the sandstone samples transitioned from splitting or low-angle shear failure to high-angle shear failure; and if the confining pressure continued to increase and exceeded the critical confining pressure of the brittle-ductile transition, a compactive strain localization failure mode with an extremely strong ductility occurred. Therefore, the failure results of the conventional triaxial tests conducted in this study are consistent with the above research, and the applied confining pressures are within the range of the transition from splitting or low-angle shear failure to high-angle shear failure.

By comparing Fig. 5.8b with a, it was found that the failure results at the macroscopic scale after the cyclic disturbance tests in triaxial stress states at the five levels of confining pressures were quite different from those after the conventional triaxial compression tests. The sandstone samples were more severely broken and fragmented after the cyclic disturbance tests than after the conventional triaxial tests. The surfaces of the main fracture fragments were broadly uneven, and there was a

Fig. 5.8 Failure modes after the experiments

great deal of spalling, chipping, and fragmentation near the main fractures. This fragmented failure is attributed to weak brittleness to some extent, indicating that cyclic disturbance weakens the brittleness of the failure of the rock samples. In addition, as can also be seen from Fig. 5.8b, as the confining pressure increased, after cyclic disturbance tests the rock samples exhibited more severe flaking and exfoliation. The main pieces of the fractured rock samples were smaller, and the number of spalling, flaking blocks increase. At a confining pressure of 40 MPa, the failure result of the sandstone sample was complete cataclasis, with no obvious main pieces. Because of the non-transparent metal confining pressure chamber, it is hard to observed the progressive damage to failure of rock samples directly, but the mechanism of this process can be can be inferred from previous experimental studies and numerical simulation results. To a certain extent, the confining pressure constrains the development of tension and shear fractures in the rock sample (Yao et al. 2016). When the rock sample is constrained by the confining pressure in the lateral direction and is subjected to continuous cyclic disturbance in the axial direction, many mesoscopic tensile and shear cracks are generated inside the sample (Zhou et al. 2020). The tips of these cracks are also constrained to develop freely, giving rise to the development of wing-shaped cracks at a certain angle (usually a small acute angle) to the principal stress (Yang and Jing 2011). Under the continuous action of the load, the tension cracks and the developing wing-shaped cracks intersect each other, influence each other, and become distributed throughout a large portion of the sample, leading to the fragmentation or even complete cataclasis of the rock sample when failure occurs under confining pressure, axial load, and axial cyclic disturbance conditions.

The entire damage and failure process, such as the initiation, propagation, connection, and nucleation of cracks, may leave some traces on the fracture surfaces, so the fracture morphology can reflect the fracture processes of the rock sample to a certain extent (Mindess and Diamond 1992; Ghamgosar and Erarslan 2016; He et al. 2019). As is shown in Fig. 5.9, using the Desktop Scanning Electron Microscope Phenom ProX© designed by Phenom-word BV Co. Ltd., a fragment that peeled off from a sandstone sample (No. RS-12-2) after the cyclic disturbance test was scanned under different magnifications ranging from 235 to 5000 times. At 235 times magnification, grain boundaries are indefinite and many microfractures are observed to split the fragment into many small irregularly shaped pieces. The plane morphology of the microfractures between the small rock pieces are mainly Y- and T-shaped, exhibiting the fracture characteristics of coexisting shear cracks and tension cracks. When the magnification is 1000 times, the surface of the small rock piece is observed to be separated by a crack with a smooth cleavage surface with a few cleavage steps and river drawing on facets. The fracture surfaces of the small pieces are relatively flat and smooth even at a magnification of 5000 times. It is noted that the microstructures of fractures under other levels of confining pressures after cyclic disturbance have similar SEM images. It can be speculated that although the ductility gradually increases and the brittleness gradually decreases with the increase of the confining pressure, the final ruptures of the rock samples after cyclic disturbance

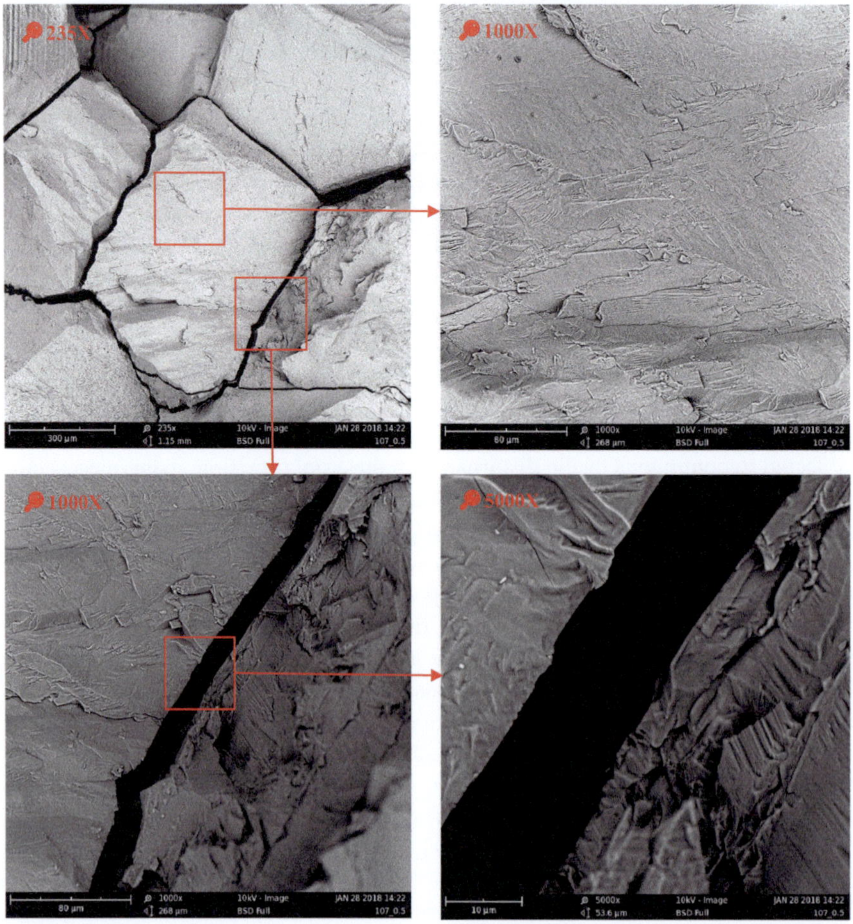

Fig. 5.9 SEM images of the fracture surfaces of rock sample debris after the triaxial cyclic dynamic disturbance test (No. RS-12-2)

under confining pressures less than or equal to 40 MPa are characteristic of intercrystalline cracking in brittle field, including shearing and cleavage fractures. In this context, dislocation slip occurs at the intercrystalline cements with low strength.

5.4 Chapter Summary

In order to gain a better understanding of the effect of the confining pressure on the damage evolution and failure behaviors of intact sandstone samples during cyclic disturbance, a series of tests with a continuous axial cyclic loading in triaxial stress states were carried out under five different levels of confining pressures ranging from 0 to 40 MPa. The following conclusions were drawn from the experimental results.

1. From the beginning of the cyclic disturbance application to the failure of the sandstone sample, both the duration and the number of cycles were positively correlated with the confining pressure. At the end of the cyclic disturbance phase, the deviatoric stress exhibited a steep instantaneous decline, indicating that the cyclic disturbance triggered the continuous accumulation of damage inside the rock samples and eventually caused the rock samples to undergo brittle failure.
2. During the cyclic disturbance, the confining pressure impeded the formation and growth of cracks in the rock samples and reduced the degree of damage during the cyclic disturbance. The increasing trend of the damage variable of the rock samples, which was defined using the maximum strain method, indicates that the damage to the rock sample developed rapidly at the beginning but relatively slowly at the end of the cyclic disturbance phase.
3. The sandstone samples were more broken and fragmented after the cyclic disturbance tests than after the conventional triaxial compression tests. As the confining pressure increased, the failure of the rock samples after the cyclic disturbance tests exhibited more severe flaking and exfoliation.

References

Bagde MN, Petroš V (2005a) Fatigue properties of intact sandstone samples subjected to dynamic uniaxial cyclical loading. Int J Rock Mech Min Sci 42(2):237–250. https://doi.org/10.1016/j.ijrmms.2004.08.008

Bagde MN, Petroš V (2005b) Waveform effect on fatigue properties of intact sandstone in uniaxial cyclical loading. Rock Mech Rock Eng 38(3):169–196. https://doi.org/10.1007/s00603-005-0045-8

Baud P, Vinciguerra S, David C et al (2009) Compaction and failure in high porosity carbonates: mechanical data and microstructural observations. Pure Appl Geophys 166(5):869–898. https://doi.org/10.1007/s00024-009-0493-2

Beer FP, Johnston ER, Dewolf JT et al (2012) Mechanics of materials. McGraw-Hill, New York

Bésuelle P, Desrues J, Raynaud S (2000) Experimental characterisation of the localisation phenomenon inside a Vosges sandstone in a triaxial cell. Int J Rock Mech Min Sci 37(8):1223–1237. https://doi.org/10.1016/S1365-1609(00)00057-5

Brace WF, Paulding BW, Scholz C (1966) Dilatancy in the fracture of crystalline rocks. J Geophys Res 71(16):3939–3953. https://doi.org/10.1029/JZ071i016p03939

Cai M, Kaiser PK, Tasaka Y et al (2004) Generalized crack initiation and crack damage stress thresholds of brittle rock masses near underground excavations. Int J Rock Mech Min Sci 41(5):833–847. https://doi.org/10.1016/j.ijrmms.2004.02.001

Campbell FC (2012) Fatigue and fracture—understanding the basics. ASM International

Cao J, Chung DDL (2002) Damage evolution during freeze-thaw cycling of cement mortar, studied by electrical resistivity measurement. Cem Concr Res 32(10):1657–1661. https://doi.org/10.1016/S0008-8846(02)00856-6

Chen X, Yu J, Tang CA et al (2017) Experimental and numerical investigation of permeability evolution with damage of sandstone under triaxial compression. Rock Mech Rock Eng 50(6):1529–1549. https://doi.org/10.1007/s00603-017-1169-3

Chen C, Xu T, Heap MJ et al (2018a) Influence of unloading and loading stress cycles on the creep behavior of Darley Dale sandstone. Int J Rock Mech Min Sci 112:55–63. https://doi.org/10.1016/j.ijrmms.2018.09.002

Chen C, Xu T, Li SH (2018b) Microcrack evolution and associated deformation and strength properties of sandstone samples subjected to various strain rates. Minerals 8(6):231. https://doi.org/10.3390/min8060231

Christensen NI, Wang HF (1985) The influence of pore pressure and confining pressure on dynamic elastic properties of Berea sandstone. Geophysics 50(2):207–213. https://doi.org/10.1190/1.1441910

Du K, Tao M, Li XB et al (2016) Experimental study of slabbing and rockburst induced by true-triaxial unloading and local dynamic disturbance. Rock Mech Rock Eng 49(9):3437–3453. https://doi.org/10.1007/s00603-016-0990-4

Gatelier N, Pellet F, Loret B (2002) Mechanical damage of an anisotropic rock under cyclic triaxial tests. Int J Rock Mech Min Sci 39(3):335–354. https://doi.org/10.1016/S1365-1609(02)00029-1

Ghamgosar M, Erarslan N (2016) Experimental and numerical studies on development of fracture process zone (FPZ) in rocks under cyclic and static loadings. Rock Mech Rock Eng 49(3):893–908. https://doi.org/10.1007/s00603-015-0793-z

He C, Yang J (2018) Laboratory study of dynamic mechanical characteristic of granite subjected to confining pressure and cyclic blast loading. Latin Am J Solids Struct 15(5):e44. https://doi.org/10.1590/1679-78254424

He W, Chen K, Hayatdavoudi A et al (2019) Effects of clay content, cement and mineral composition characteristics on sandstone rock strength and deformability behaviors. J Petrol Sci Eng 176(5):962–969. https://doi.org/10.1016/j.petrol.2019.02.016

Kachanov ML, Krajcinovic D (1986) Introduction to continuum damage mechanics. J Appl Mech 54(2):481. https://doi.org/10.1007/978-94-017-1957-5

Kaiser PK, Tang CA (1998) Numerical simulation of damage accumulation and seismic energy release during brittle rock failure—Part II: Rib pillar collapse. Int J Rock Mech Min Sci 35(2):123–134. https://doi.org/10.1016/s0148-9062(97)00010-7

Kaiser PK, Yazici S, Maloney S (2001) Mining-induced stress change and consequences of stress path on excavation stability—a case study. Int J Rock Mech Min Sci 38(2):167–180. https://doi.org/10.1016/S1365-1609(00)00038-1

Li HB, Zhao J, Li TJ (1999) Triaxial compression tests on a granite at different strain rates and confining pressures. Int J Rock Mech Min Sci 36(8):1057–1063. https://doi.org/10.1016/S1365-1609(99)00120-3

Liang Y, Li Q, Gu Y et al (2017) Mechanical and acoustic emission characteristics of rock: effect of loading and unloading confining pressure at the postpeak stage. J Nat Gas Sci Eng 44:54–64. https://doi.org/10.1016/j.jngse.2017.04.012

Liu EL, He SM (2012) Effects of cyclic dynamic loading on the mechanical properties of intact rock samples under confining pressure conditions. Eng Geol 125(27):81–91. https://doi.org/10.1016/j.enggeo.2011.11.007

Liu Q, Cheng Y, Jin K et al (2017) Effect of confining pressure unloading on strength reduction of soft coal in borehole stability analysis. Environ Earth Sci 76(4):173. https://doi.org/10.1007/s12665-017-6509-9

Mindess S, Diamond S (1992) SEM investigations of fracture surfaces using stereo pairs: II. Fracture surfaces of rock-cement paste composite specimens. Cem Concr Res 22(4):678–688. https://doi.org/10.1016/0008-8846(92)90020-V

Oda M, Suzuki K, Maeshibu T (1984) Elastic compliance for rock-like materials with random cracks. Soils Found 24:27–40. https://doi.org/10.3208/sandf1972.24.3_27

Ougier-Simonin A, Fortin J, Gueguen Y et al (2011) Cracks in glass under triaxial conditions. Int J Eng Sci 49(1):105–121. https://doi.org/10.1016/j.ijengsci.2010.06.026

Paterson MS (1958) Experimental deformation and faulting in Wombeyan marble. Geol Soc Am Bull 69(4):465–476. https://doi.org/10.1130/0016-7606(1958)69[465:EDAFIW]2.0.CO;2

Paterson MS, Wong TF (2005) Experimental rock deformation—the brittle field. Springer, Berlin

Peng K, Zhou J, Zou Q et al (2019) Deformation characteristics of sandstones during cyclic loading and unloading with varying lower limits of stress under different confining pressures. Int J Fatigue 127(10):82–100. https://doi.org/10.1016/j.ijfatigue.2019.06.007

Rabczuk T, Zi G, Bordas S et al (2010) A simple and robust three-dimensional cracking-particle method without enrichment. Comput Methods Appl Mech Eng 199(37–40):2437–2455. https://doi.org/10.1016/j.cma.2010.03.031

Schijve J (2009) Fatigue of structures and materials, 2nd edn. Springer, Dordrecht

Schöpfer MPJ, Abe S, Childs C et al (2009) The impact of porosity and crack density on the elasticity, strength and friction of cohesive granular materials: insights from DEM modelling. Int J Rock Mech Min Sci 46(2):250–261. https://doi.org/10.1016/j.ijrmms.2008.03.009

Simpson C (1985) Deformation of granitic rocks across the brittle-ductile transition. J Struct Geol 7(5):503–511. https://doi.org/10.1016/0191-8141(85)90023-9

Sun B, Zhu Z, Shi C et al (2017) Dynamic mechanical behavior and fatigue damage evolution of sandstone under cyclic loading. Int J Rock Mech Min Sci 94:82–89. https://doi.org/10.1016/j.ijrmms.2017.03.003

Takemura T, Oda M (2005) Changes in crack density and wave velocity in association with crack growth in triaxial tests of Inada granite. J Geophys Res Atmos 110(B5):1–14. https://doi.org/10.1029/2004JB003395

von Kármán T (1911) Festigkeitsversuche unter allseitigem Druck. Z Verein deutsch Ing 55:1749–1757 (in Germany)

Wong TF (1982) Micromechanics of faulting in westerly granite. Int J Rock Mech Min Sci Geomech Abstr 19(2):49–64. https://doi.org/10.1016/0148-9062(82)91631-X

Wong TF, Baud P (2012) The brittle-ductile transition in porous rock: a review. J Struct Geol 44:25–53. https://doi.org/10.1016/j.jsg.2012.07.010

Wong LNY, Li D, Liu G (2013) Experimental Studies on permeability of intact and singly jointed meta-sedimentary rocks under confining pressure. Rock Mech Rock Eng 46(1):107–121. https://doi.org/10.1007/s00603-012-0251-0

Xiao JQ, Ding DX, Jiang FL et al (2010) Fatigue damage variable and evolution of rock subjected to cyclic loading. Int J Rock Mech Min Sci 47:461–468. https://doi.org/10.1016/j.ijrmms.2009.11.003

Yang SQ, Jing HW (2011) Strength failure and crack coalescence behavior of brittle sandstone samples containing a single fissure under uniaxial compression. Int J Fract 168(2):227–250. https://doi.org/10.1007/s10704-010-9576-4

Yao M, Rong G, Zhou C et al (2016) Effects of thermal damage and confining pressure on the mechanical properties of coarse marble. Rock Mech Rock Eng 49(6):2043–2054. https://doi.org/10.1007/s00603-016-0916-1

Zhang CL, Jiang GL, Su LJ et al (2018) Dynamic behaviour of weathered red mudstone in Sichuan (China) under triaxial cyclic loading. J Mt Sci 15:1789–1806. https://doi.org/10.1007/s11629-017-4756-6

Zhou Y, Zhao D, Tang Q et al (2020) Experimental and numerical investigation of the fatigue behaviour and crack evolution mechanism of granite under ultra-high-frequency loading. R Soc Open Sci 7(4):200091. https://doi.org/10.1098/rsos.200091

Zhu WC, Li ZH, Zhu L et al (2010) Numerical simulation on rockburst of underground opening triggered by dynamic disturbance. Tunn Undergr Space Technol 25(5):587–599. https://doi.org/10.1016/j.tust.2010.04.004

Zhu HH, Zhou S, Yan ZG et al (2016) A two-dimensional micromechanical damage-healing model on microcrack-induced damage for microcapsule-enabled self-healing cementitious composites under compressive loading. Int J Damage Mech 25(5):727–749. https://doi.org/10.1177/1056789516641593

Chapter 6
Mechanism of Damage and Failure of Rock Triggered by Cyclic Dynamic Disturbance

6.1 Introduction

In the study of rock mechanics, it is generally believed that the failure of rock materials is the result of the damage process of micropores and microcracks in the rock that gradually germinate, expand, aggregate and nucleate under external loads, and finally form macroscopic fracture surfaces. The main theories used in these studies of related mechanisms are the theory of failure mechanics, that is, the theory of damage mechanics and fracture mechanics. The research of damage mechanics mainly focuses on the phenomenon that macroscopic invisible micro-defects of materials under external loads (including monotone loads, dynamic loads and cyclic loads, etc.) lead to the weakening of internal cohesion of materials and the failure of mesoscopic volume units (Chang and Lee 2004; Chen et al. 2018; Zheng et al. 2024), and the fracture mechanics is mainly concerned with the deformation properties and crack propagation laws of objects containing macroscopic visible cracks under external loads (Aliha et al. 2013; Yin et al. 2018). The failure mechanics combined with damage mechanics and fracture mechanics forms the basis for studying the failure mechanism of rocks.

In the abovementioned experiments on damage and failure of rock triggered by dynamic disturbance, random secondary cracks are generated in the rock during the quasi-static loading phase, and microcracks gradually nucleate in the dominant direction in the cyclic disturbance phase, eventually leading to abrupt instability failure of the rock sample. Therefore, this chapter attempts to explain the damage and failure mechanism of rocks under high stress coupling with dynamic disturbance using theories of failure mechanics. The main goals of the study in this chapter are as follows:

1. Establishing the statistical damage constitutive equation of rocks under quasi-static loading stage, and analyzing the damage evolution mechanism in the cyclic disturbance phase.

X. Chen, *Damage and Failure of Rock Triggered by Dynamic Disturbance*,
https://doi.org/10.1007/978-981-96-9486-0_6

2. Explaining the abrupt instability of rock samples triggered by cyclic dynamic disturbance using the cusp catastrophic theory.

6.2 Mechanism of Progressive Damage of Rock Triggered by Cyclic Dynamic Disturbance

6.2.1 Damage Constitutive Equation of Rock During Quasi-static Loading Phase

Defining damage variables, establishing damage evolution equation, and establishing the constitutive relationship of damaged rocks are the three core problems in describing the damage evolution process of rocks. Therefore, these three problems are also emphasized in carrying out the damage evolution mechanism of rocks during the quasi-static loading phase. In the quasi-static loading stage of the previous tests, the axial load is increased from zero to a certain stress level (85% of the triaxial (or uniaxial) compressive strength). In this process, the damage of rock samples is mainly due to the initiation and propagation of randomly distributed primary and secondary microcracks in rock samples, but the nucleation of microfractures has not yet occurred. The sandstone samples used in the tests have no obvious bedding, and the consistency between the samples is ensured as much as possible during the preparation and screening process, which determines that the damage distribution of rock samples during the triaxial loading process should be uniform and random in the entire volume space of rock samples. The influence of these uniform and random damages on the overall mechanical properties of rock does not have significant differences in all directions. Therefore, it can be assumed that the damage evolution of rocks during quasi-static loading is isotropic (Chai 2009), and the damage degree of rocks can be characterized by a scalar.

Lemaitre J. et al. (Lemaitre and Desmorat 2005; Lemaitre 1985) proposed the concept of effective stress and the related theory of the strain equivalence principle when defining the one-dimensional surface damage variables, which is defined as:

$$\tilde{\sigma} = \frac{F}{S - S_D} \tag{6.1}$$

where $\tilde{\sigma}$ is the effective stress, F is the normal external load acting on the damaged body, S is the cross-sectional area of the damaged body, S_D is the sum of the areas of all defects on the cross section, and $S - S_D$ is the actual area of the load.

The principle of strain equivalence is that any strain constitutive equation established for a damaged material can be derived in the same way as for a non-damaged material, simply by replacing the stress in it with the effective stress (Lemaitre and Desmorat 2005).

Starting from these two definitions and combined with the concept of micromechanics, one-dimensional surface damage variable D can be defined as the ratio of the number of damaged micro-elements to the total number of micro-elements. The scale of micro-elements should be large enough (including a certain number of microscopic cracks) and small enough (to be able to use continuum damage mechanics for analysis). The mathematical relationship between the effective stress $\tilde{\boldsymbol{\sigma}}$ and the nominal stress $\boldsymbol{\sigma}$ can then be established:

$$\tilde{\sigma} = \frac{\sigma}{1 - D} \tag{6.2}$$

In the quasi-static loading stage of this study, although the rock samples are under three-dimensional loading, all the cores are right cylinder, and the direction of lateral confining pressure is directed towards the axis of the cylinder. Therefore, the above one-dimensional relationship (Eq. 6.2) can be extended to the three-dimensional problem of this study.

Before establishing the damage evolution equation of quasi-static loading stage, three basic assumptions are made:

(1) The micro-elements with microscopic defects in the rock all satisfy the generalized Hooke's law.
(2) The failure conditions of the rock microbodies conform to the Hoek–Brown strength criterion (Hoek 1983; Hoek and Brown 1997, 2019), which is expressed as follows:

$$\sigma_1 = \sigma_2 + \sigma_c \left(m_b \frac{\sigma_3}{\sigma_c} + s \right)^{\alpha'} \tag{6.3}$$

where $\boldsymbol{\sigma}_1$, $\boldsymbol{\sigma}_2$, and $\boldsymbol{\sigma}_3$ are the maximum, middle, and minimum principal stresses, respectively; $\boldsymbol{\sigma}_c$ is uniaxial compressive strength; m_b is the value of empirical parameter m, and its value is mainly based on the geological strength index (GSI); s and $\boldsymbol{\alpha}'$ are the rock mass material constants, and $s = 1$ and $\boldsymbol{\alpha}' = 0.5$ for intact rock.

(3) The strength of rock microbodies follows Weibull distribution (Tang 1997; Tang et al. 2000a, b), and its probability density function is expressed as follows:

$$\varphi(\alpha) = \frac{n}{\alpha_0} \cdot \left(\frac{\alpha}{\alpha_0} \right)^{n-1} \cdot e^{-\left(\frac{\alpha}{\alpha_0}\right)^n} \tag{6.4}$$

where $\boldsymbol{\alpha}$ is the strength of a rock microbody; $\boldsymbol{\alpha}_0$ and n are the scale parameter and shape parameter of the Weibull distribution. Figure 6.1 shows the probability density function curves of Weibull distribution when a certain $\boldsymbol{\alpha}_0$ is selected and different shape parameter n is taken.

If the probability of damage of the rock microbody is a function of the probability density of its intensity $F = f(\tilde{\sigma})$, the relationship between the damage variable D and the damage probability density of the rock microbody $\boldsymbol{\varphi}(\boldsymbol{\alpha})$ can be expressed as:

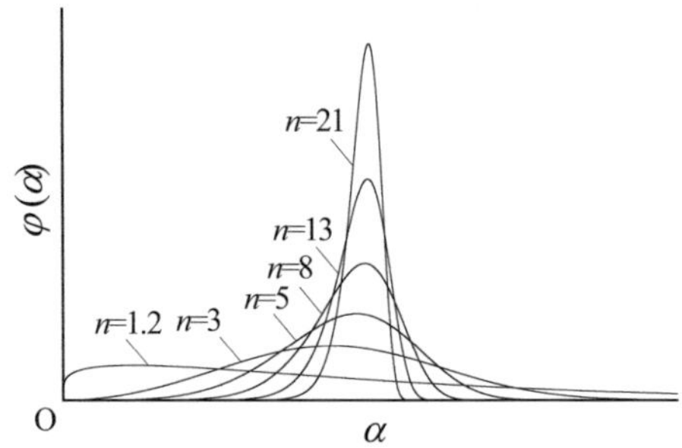

Fig. 6.1 Probability density functional images of Weibull distribution with different shape parameters *n*

$$D = \int_0^{f(\tilde{\sigma})} \varphi[f(\tilde{\sigma})]dx \tag{6.5}$$

By substituting Eq. (6.4) into Eq. (6.5), we get

$$D = 1 - \exp\left[-\left(\frac{\alpha}{\alpha_0}\right)^n\right] \tag{6.6}$$

In damage mechanics, the elastic modulus of the damaged material is defined as (Lemaitre and Desmorat 2005)

$$\tilde{E} = E(1 - D) = E\exp\left[-\left(\frac{\alpha}{\alpha_0}\right)^n\right] \tag{6.7}$$

By substituting Eq. (6.7) into the generalized Hooke's law, the damage constitutive relation based on probability distribution can be obtained in the quasi-static loading stage of rock microbodies

$$\sigma_1 = \varepsilon_1\tilde{E} + \nu(\sigma_2 + \sigma_3) = \varepsilon_1 E\exp\left[-\left(\frac{\alpha}{\alpha_0}\right)^n\right] + \nu(\sigma_2 + \sigma_3) \tag{6.8}$$

where $\boldsymbol{\sigma}_1$ represents the stress on the rock microbody along the axial direction of the macroscopic rock sample. $\boldsymbol{\sigma}_2$ and $\boldsymbol{\sigma}_3$ denote the radial and tangential stresses along the macroscopic rock sample on the rock microbody, respectively. In conventional triaxial tests, $\boldsymbol{\sigma}_2 = \boldsymbol{\sigma}_3$ is equal to the confining pressure applied to the rock sample. $\boldsymbol{\varepsilon}_1$ is the strain along the macroscopic sample axis on the rock element, and $\boldsymbol{\nu}$ is Poisson's ratio.

Zheng (2002) introduced the concept of effective stress into Eq. (6.3) and obtained the expression:

$$f(\tilde{\sigma}) = \frac{1}{3}m\sigma_{1c}\tilde{I}_1 + \cos^2\theta_\sigma + m\sigma_{1c}\sqrt{\tilde{J}_2}\left(\cos\theta_\sigma + \frac{1}{3}\sin\theta_\sigma\right) = s\sigma_{1c}^2 \tag{6.9}$$

$$\tilde{I}_1 = \tilde{\sigma}_1 + \tilde{\sigma}_2 + \tilde{\sigma}_3 = \frac{E\varepsilon_1(\sigma_1 + \sigma_2 + \sigma_3)}{\sigma_1 - \nu(\sigma_2 + \sigma_3)} \tag{6.10}$$

$$\begin{aligned}\tilde{J}_2 &= \frac{1}{6}\left[(\tilde{\sigma}_1 - \tilde{\sigma}_2)^2 + (\tilde{\sigma}_2 - \tilde{\sigma}_3)^2 + (\tilde{\sigma}_3 - \tilde{\sigma}_1)^2\right] \\ &= \frac{E^2\varepsilon_1^2\left(\sigma_1^2 + \sigma_2^2 + \sigma_3^2 - \sigma_1\sigma_2 - \sigma_2\sigma_3 - \sigma_3\sigma_1\right)}{3[\sigma_1 - \nu(\sigma_2 + \sigma_3)]^2}\end{aligned} \tag{6.11}$$

where $\tilde{I}_1$ and $\tilde{J}_2$ are the first invariants of the effective stress and the second invariant of the effective stress deviation, respectively. m and s are empirical parameters related to the degree of hardness and fragmentation of rocks, respectively. The value of m ranges from 1×10^{-7} to 25. When m is taken as 25, the rock is complete and hard, and when m is taken as 1×10^{-7}, the rock mass has a history of serious disturbance. The value of s ranges from 0 to 1. When s is take as 1, the rock is very intact, and when s is take as 0, the rock is very broken. $\boldsymbol{\theta}_\sigma$ is the Lorde angle, which is usually taken as 30° in conventional triaxial loading.

The Hoek–Brown criterion is satisfied when the microbodies are damaged, therefore Eqs. (6.9), (6.10) and (6.11) are substituted into Eq. (6.8) to obtain the damage constitutive relationship based on Weibull distribution of rock samples during conventional triaxial compression tests:

$$\sigma_1 = \varepsilon_1 E \exp\left\{-\left[\frac{\frac{1}{3}m\sigma_{1c}\tilde{I}_1 + \cos^2\theta_\sigma + m\sigma_{1c}\sqrt{\tilde{J}_2}\left(\cos\theta_\sigma + \frac{1}{3}\sin\theta_\sigma\right)}{\alpha_0}\right]^n\right\} + \nu(\sigma_2 + \sigma_3) \tag{6.12}$$

Further, we get:

$$\sigma_1 = \varepsilon_1 E \exp\left[-\left(\frac{m\sigma_c\sigma_1 E\varepsilon_1}{\alpha_0\sigma_1 - 2\nu\alpha_0\sigma_3} + \frac{\sigma_1^2E^2\varepsilon_1^2 - 2\sigma_1\sigma_3E^2\varepsilon_1^2 + \sigma_3^2E^2\varepsilon_1^2}{\alpha_0\sigma_1^2 - 4\nu\alpha_0\sigma_1\sigma_3 + 4\nu^2\alpha_0\sigma_3^2}\right)^n\right] + 2\nu\sigma_3 \tag{6.13}$$

Figure 6.2 shows a typical stress–strain curve of rock under conventional triaxial compression, from which some characteristics can be summarized as boundary conditions of Eq. (6.13):

(1) The curve passes through the coordinate origin, i.e., when $\boldsymbol{\varepsilon}_1 = 0$, $\boldsymbol{\sigma}_1 = 0$.
(2) The peak point of the curve can be determined by experiment, i.e., when $\boldsymbol{\varepsilon}_1 = \boldsymbol{\varepsilon}_{1c}$, $\boldsymbol{\sigma}_1 = \boldsymbol{\sigma}_{1c}$.
(3) The slope of the curve at the peak point is zero, i.e., when $\boldsymbol{\varepsilon}_1 = \boldsymbol{\varepsilon}_{1c}$, $\frac{d\sigma_1}{d\boldsymbol{\varepsilon}_1} = 0$.

In this study, conditions (2) and (3) are used to establish equations.

Among the nine parameters of Eq. (6.13), $\boldsymbol{\sigma}_3$ is the confining pressure condition given in the experiment, $\boldsymbol{\sigma}_1$, $\boldsymbol{\sigma}_c$, $\boldsymbol{\varepsilon}_c$, E and $\boldsymbol{\nu}$ can be directly measured or calculated through the experimental result. For a complete rock sample, m is directly taken as 25.

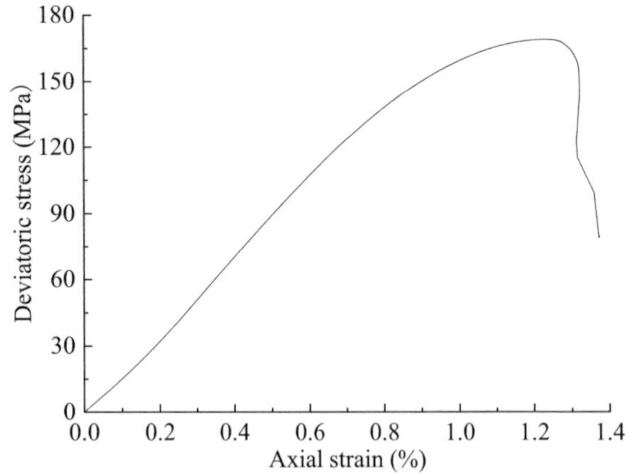

Fig. 6.2 Typical stress–strain curves of rock under conventional triaxial compression (obtained from the tests under the confining pressure of 30 MPa)

Thus, the only unknown parameters left in Eq. (6.13) are n and $\boldsymbol{\alpha}_0$. Using the above three conditions, the unknown parameters n and $\boldsymbol{\alpha}_0$ can be obtained by inputting the characteristic parameters measured under confining pressure in Table 6.1 into a MATLAB program. The calculation results are listed in Table 6.2.

By taking the mean values of unknown parameters n and $\boldsymbol{\alpha}_0$ obtained under five levels of confining pressures as the final values, and substituting the characteristic parameters in Table 6.1 into Eq. (6.13), the stress–strain curve under each confining pressure can be calculated. It is worth noting that in the triaxial cases, the axial stress

Table 6.1 The characteristic parameters measured under each confining pressure

Confining pressure $\boldsymbol{\sigma}_3$/MPa	Peak stress σ_{1c}/MPa	Axial strain at peak stress ε_{1c}/%	Elastic modulus E/GPa	Poisson's ratio ν
0	60.45	0.526	10.826	0.214
10	109.17	0.942	11.020	0.253
20	139.12	1.022	13.540	0.329
30	168.99	1.230	13.178	0.366
40	199.25	1.351	15.111	0.388

Table 6.2 Unknown parameters in Eq. (6.13) solved by MATLAB programming

Confining pressure σ_3/MPa	0	10	20	30	40	Mean
Unknown parameter n	9.414	10.308	9.873	9.275	9.639	9.7018
Unknown parameter $\boldsymbol{\alpha}_0$	538,475.9	812,380.5	981,441.5	1,110,693.1	142,878.3	1,823,173.86

σ_1 on the left side of the equal sign of Eq. (6.13) is replaced by the deviatoric stress $\sigma_1 - \sigma_3$. Figure 6.3 illustrates the comparison diagram between the theoretical curves and the experimental curves under confining pressure ranging from 0 to 40 MPa. As can be seen from the figure, the theoretical curves are roughly consistent with the experimental curves. Since the rock constitutive relationship proposed in this study is derived based on the peak stress σ_{1c}, axial strain at peak stress ε_{1c}, etc., the theoretical curves at the peak points are in complete agreement with the experimental curves. In the actual application process, data such as peak stress, axial strain at peak stress, elastic modulus and Poisson's ratio of various rocks under various confining pressures can be obtained through a large number of triaxial tests, and their variation laws with confining pressures can be summarized. By substituting these laws into Eq. (6.13), the proposed constitutive relationship can be extended to other types of rock materials.

In addition, the stress–strain curve under triaxial compression can be divided into different stages using total measured volumetric strain and calculated crack volumetric strain, including crack closure stage, linear elastic deformation stage, steady crack propagation stage, yield stage, and post-peak stage (Martin 1997; Eberhardt et al. 1998; Cai et al. 2004; Chen et al. 2017). According to this division method, the quasi-static loading process in these tests is within 85% of the peak stress, which mainly includes t crack closure stage, linear elastic deformation stage, and steady crack propagation stage. For the sake of simplicity, it is sufficient to fit the curves directly as straight lines, but this does not characterize the strain softening stage of the rock after the peak. The constitutive relationship of the rock proposed in this study (Eq. 6.13) not only takes into account the quasi-static loading stage, but also reflects yield stage and post-peak stage, which is more applicable and more in line with the characteristics of rock materials than simply using straight lines as constitutive conditions.

From Eq. (6.13), we can also derive the expression for the damage variable D:

$$D = 1 - \exp\left[-\left(\frac{m\sigma_c\sigma_1 E\varepsilon_1}{\alpha_0\sigma_1 - 2\nu\alpha_0\sigma_3} + \frac{\sigma_1^2E^2\varepsilon_1^2 - 2\sigma_1\sigma_3E^2\varepsilon_1^2 + \sigma_3^2E^2\varepsilon_1^2}{\alpha_0\sigma_1^2 - 4\nu\alpha_0\sigma_1\sigma_3 + 4\nu^2\alpha_0\sigma_3^2}\right)^n\right] \tag{6.14}$$

Correspondingly, the evolution of the damage variable D under each confining pressure can be calculated, and the relationship between the damage variable D and the axial strain ε_1 is plotted in Fig. 6.4. It can be seen from the figure that under the five confining pressures, the evolution trend of the damage variables along with the axial strain shows an "S" shape in general. At the initial stage of loading, the damage of the rock sample is relatively slight, and the damage variable is close to zero. After loading to a certain extent, the damage variable increases exponentially. Subsequently, the growth rate of damage variable becomes smaller again and finally reaches 1, that is, macroscopic rupture is formed in the rock sample. From the point of view of the shape of the curve, these "S" shaped curves are very in line with the characteristics of each stage of stress–strain curve of the rock under conventional triaxial compression. In addition, the figure also shows that the larger the confining pressure, the greater

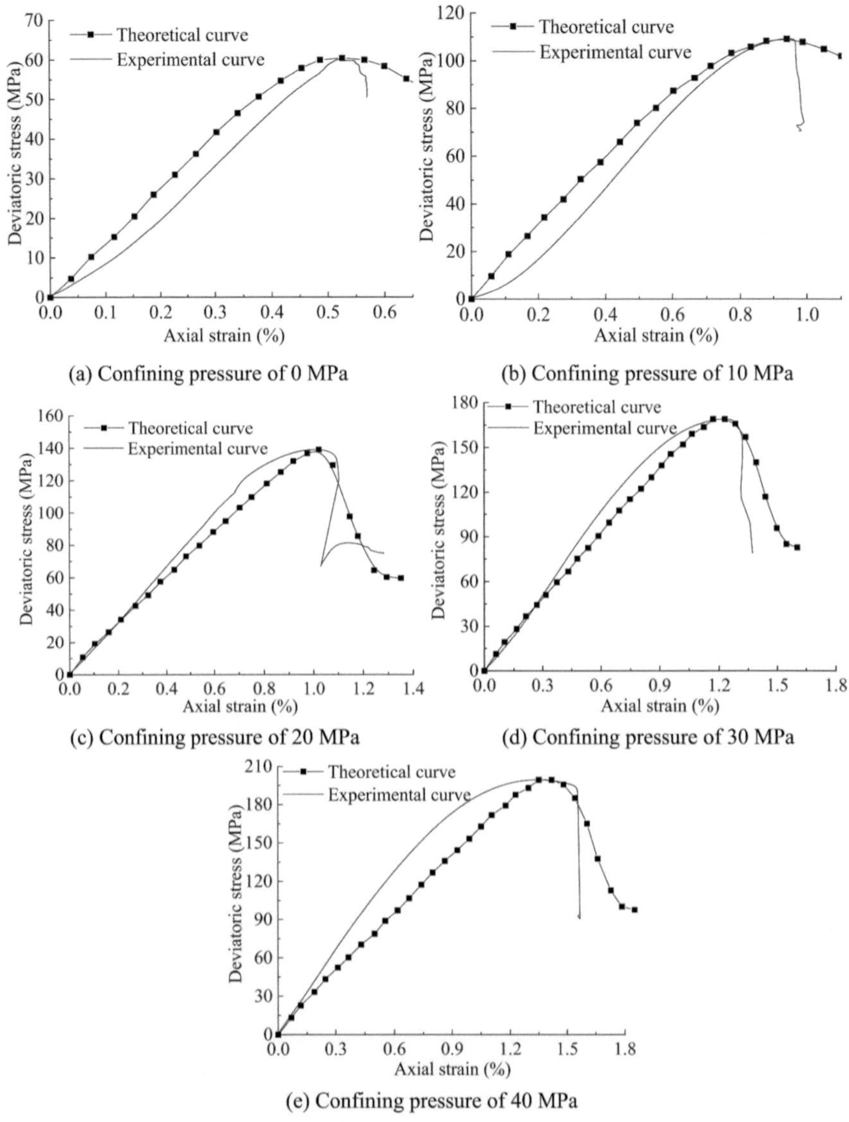

Fig. 6.3 Comparison of stress–strain curves of sandstone between experimental results and theoretical results based on statistical damage constitutive model

the axial strain experienced by increasing the damage variable from 0 to 1, which reflects that the larger the confining pressure, the stronger the ability of the rock sample to resist deformation, and it is also consistent with the characteristics of the rock. Therefore, it is reasonable to take Eq. (6.13) as the constitutive relationship and Eq. (6.14) as the evolution of the cumulative damage of the sandstone used in

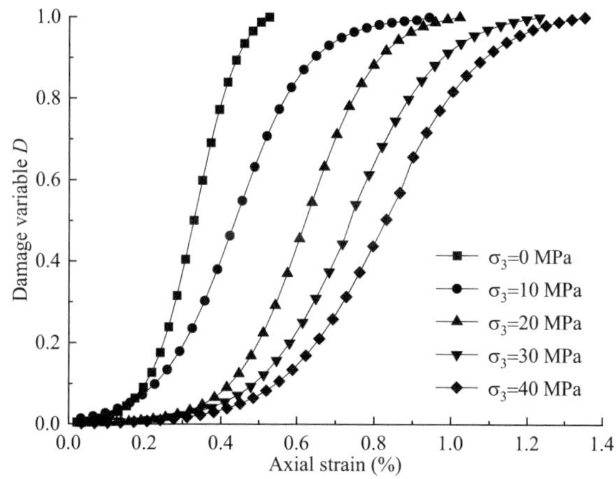

Fig. 6.4 Theoretical models of damage variable with axial strain of sandstone under five different confining pressures

this study under the quasi-static loading conditions. In Eq. (6.14), $\boldsymbol{\varepsilon}_1$ takes the axial strain corresponding to 85% of the peak stress, and the damage of the rock sample at the end of the quasi-static loading phase can also be obtained, which corresponds to the initial damage in the cyclic dynamic disturbance phase, and also satisfies the four basic requirements for defining the damage variables proposed by Xiao et al. (2010).

6.2.2 Damage Evolution Mechanism of Rock During Cyclic Dynamic Disturbance Loading Stage

In Sects. 4.3.3 and 5.3.3, it has been expounded in detail the rationality of using the concepts of residual strain, average dynamic axial stiffness and maximum strain to define the damage variable D in the cyclic dynamic disturbance loading stage. If the relationship between the damage variable D and the relative cycles n/N is fitted, the empirical function of the cumulative damage evolution law in the cyclic dynamic disturbance loading stage can also be obtained. To avoid repetition, it will not be discussed in this section. However, the mechanical mechanism of the rock damage and degradation process in the cyclic dynamic disturbance loading stage has not been given in the previous sections, and the relevant mechanism will be explained in this section.

In the cyclic dynamic disturbance loading stage of the relevant experiments in this study, the rock is subjected to both constant static load and periodic cyclic dynamic disturbance, so the damage evolution process must be the result of the combination of the two. First, the static load action on the rock samples at the cyclic dynamic disturbance loading stage is analyzed. If the axial pressure is kept at the level of 85% of the peak stress with no cyclic dynamic disturbance while the confining pressure is kept constant, the axial strain of a rock sample will gradually increase with the

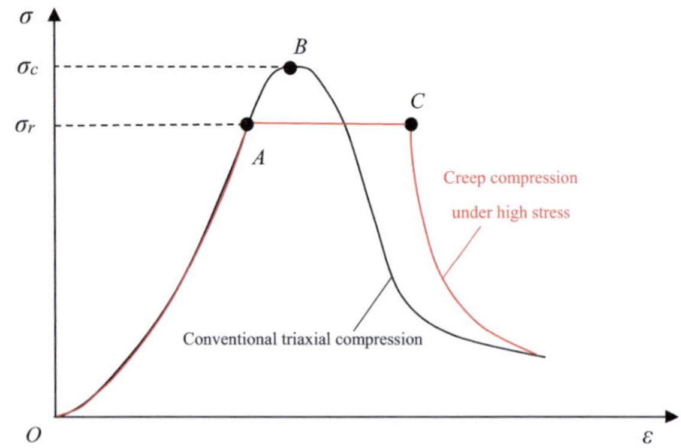

Fig. 6.5 Schematic diagram for stress path of creep test with high stress level (Yang et al. 2011)

increase of time. Yang et al. (2011) and Su et al. (2016) conducted a creep test under high stress levels (stress level σ_r is slightly lower than peak stress σ_c), also known as "time-lag test", which is different from conventional uniaxial compression test (as shown in Fig. 6.5). In the cyclic dynamic disturbance loading stage in this study, it can actually be regarded as a static load and a cyclic disturbance load superimposed on the rock samples at a high stress level, and the process of static load is actually similar to the creep compression test at high stress level. Yang et al. (2011) believe that the development rate of radial strain of rock samples is greater than that of axial strain during creep loading at high stress levels. Compared with the conventional triaxial compression test, the radial strain development rate of the rock is also relatively fast. After the creep test at high stress level, the rock samples usually form more flaky debris, which is not caused by the slip dislocation of rock mineral crystals, but by the splitting and opening of vertical cracks.

In the creep damage evolution process under high stress, one or several dominant cracks in the rock usually formed in the process of quasi-static loading propagate at a slower rate than that of conventional triaxial compression. These cracks are at a stress level of 85% of the peak strength and no longer increase, not reaching the stress state of rock failure, and the crack propagation rate is low and the time it takes is longer. The dominant cracks are generated in the quasi-static stage, results in a random distribution in the location and direction of the rock samples (in general, it does not coincide with the direction of the principal stress). While, in the creep damage process under high stress, these cracks will open radially to the rock sample and continue to propagate, so that the failure form of flake clastic finally form, which is also in line with the characteristics of the rock sample after cyclic dynamic disturbance in this test.

From the test, the amplitude of the cyclic disturbance test does not reach 15% of the peak stress of the rock samples, that is, the stress state of the rock samples is smaller than the stress state at the peak of the conventional load even when the disturbance load reaches the peak. Therefore, the rock under cyclic disturbance load does not cause damage and failure because the stress is close to its failure stress. In fact, the stiffness of the rock material deteriorates when the cyclic dynamic disturbance load acts on high stress conditions compared with when it is not subjected to the cyclic disturbance load. In addition, the utilization of the energy received during the cyclic dynamic disturbance process is optimized. The deterioration of stiffness is mainly reflected in the fact that under the action of cyclic dynamic disturbance load, the cracks generated by the rock samples in the quasi-static loading stage continue to germinate and propagate, and the damage continues to accumulate, resulting in the slow dissipation of the strain energy received by the rock samples in the process of damage evolution. The improvement of energy utilization rate is mainly reflected in that, because the rock samples are in a constant state of confining pressure and axial pressure, some discontinuity such as primary and secondary cracks inside the rock samples are just in a closed continuous state under the action of static load, which brings about many cracks that stop expanding after originally expanding to the discontinuous interface and continue to expand through the closed interface. The utilization rate of the received disturbance energy of the rock samples is improved, and the number of cracks inside the rock samples is increased, which also leads to the disintegration of the residual sample during the final failure.

6.3 Mechanism of Abrupt Instability of Rock Triggered by Cyclic Dynamic Disturbance

6.3.1 Microcrack Nucleation in Rock During Cyclic Dynamic Disturbance Phase

From the results of cyclic dynamic disturbance experiments in Chaps. 4 and 5, it can be seen that the rock samples will suddenly destabilize to a certain extent under the action of cyclic dynamic disturbance, that is, the bearing capacity decreases rapidly, and the deviatoric stress–time curves fall for a short time, showing typical brittle failure characteristics. From the perspective of mesoscopic failure mechanics, it is generally believed that the process of damage evolution of concrete and rock materials eventually leading to failure is actually the process of continuous nucleation of microscopic cracks, that is, under the action of load, the micro-fractures near the crack tips of primary cracks and secondary cracks continue to form and propagate, and finally develop to the extent of crossing, penetrating, interacting, and even merging. The region where the micro-crack develops and evolves near the primary crack and secondary crack is also called the process region (Atkinson and Meredith 1987). From a macroscopic point of view, various strength or failure criteria in rock mechanics, such as Mohr–Coulomb criterion (Joseph and Arno 2012), Hoek–Brown criterion (Hoek 1983; Hoek and Brown 1997, 2019), Druker-Prager criterion (Alejano and Bobet 2012), unified strength criterion (Yu et al. 2002, 2011 Yu 2004), etc., are actually macroscopic statistical results of the development, evolution and nucleation process of microfracture in the process of damage evolution.

As mentioned above, in the triaxial cyclic dynamic disturbance tests in this study, the quasi-static loading stage ends at about 85% of the peak strength, roughly corresponding to the process of random initiation, development and expansion of rock microcracks. After entering the cyclic dynamic disturbance loading stage, the evolution process of microcracks is transformed into the process of aggregation nucleation, that is, the microcrack has a certain advantageous direction. Under the action of cyclic dynamic disturbance, the relatively dense microcracks in the rock samples will interact with each other, causing some of them to close, while another part of them to newly form. These newly generated microcracks gradually nucleate in the dominant direction, resulting in an obvious trend in the arrangement of microcracks inside the rock samples. When the density of microcracks in the dominant direction reaches a critical or threshold value, the continuously applied cyclic dynamic disturbance is equivalent to a high frequency impact load at this moment, which will cause the internal cracks to break through instantaneously and the rock samples to lose their bearing capacity. In the process of nucleation of microcracks with a certain direction, it will also promote the heterogeneity and anisotropy of the rock samples, and further lead to the uneven distribution of stress field in the interior, which also accelerates the nucleation of microcracks and the formation of macroscopic fracture to a certain extent, and shortens the life of damage process triggered by the cyclic dynamic disturbance. It is worth noting that previous studies believed that the sudden drop of the stress–strain curves at the peak in the quasi-static loading tests is caused by insufficient stiffness of the testing machine (Zhou 1999). But, according to the stress–strain curves obtained from the conventional loading test (as shown in Fig. 3. 4), the stiffness of the testing machine used in this study is sufficient. In addition, the cyclic dynamic disturbance also reduces the stiffness of the rock, that is, the relative stiffness of the testing machine and the rock sample is increased. Therefore, the stress drop at the end of the cyclic dynamic disturbance stage is unlikely to be caused by insufficient stiffness of the testing machine, but more likely to be caused by the nucleation of microcracks.

From Figs. 4.2 and 5.2, the duration of microcrack nucleation is related to the frequency and amplitude of the cyclic dynamic disturbance loads. Kuksenko et al. (1996) gave an empirical formula for the nucleation life of rocks under uniaxial tensile conditions:

$$\tau=\tau_0 \exp\left[\frac{(U_0-\gamma\sigma)}{kT}\right] \tag{6.15}$$

where $\boldsymbol{\tau}$ is the nucleation life, $\boldsymbol{\tau}_0$ is the period of self-excited vibration of atoms in a solid, U_0 is fracture activity energy, $\boldsymbol{\gamma}$ is the parameter related to the mechanical properties of materials, $\boldsymbol{\sigma}$ is the applied stress, k is the Boltzmann constant, and T is the thermodynamic temperature. Assuming that the material does not exchange heat with the surrounding environment during the loading process, that is, the temperature does not change, it can be concluded from the equation that the smaller the load, the longer the nucleation life of the sample. Conversely, the greater the load, the shorter the nucleation life. This empirical formula is also applicable to nucleation process of microcracks in the rock samples subjected to the cyclic dynamic disturbance. The higher the frequency and the greater the amplitude of the cyclic dynamic disturbance, the shorter the nucleation lifetime, and vice versa.

6.3.2 *Analytical Analysis of the Sudden Stress Drop at the End of the Cyclic Disturbance Phase Using Catastrophe Theory*

6.3.2.1 Brief Description of Cusp Catastrophe Theory

Catastrophe theory was established based on mathematical theories, such as the topology and singularity theory, which is used to build mathematical models for various catastrophe phenomena. As was described by Saunders (1980), a system that has smooth behaviors at most times and in most places may encounter discontinuities at some special time, points, or places. It is assumed that the state of the system at any time can be fully specified by providing the values of n variables $(x_1, x_2, \ldots, x_n)$ that are determined by m independent variables $(u_1, u_2, \ldots, u_n)$. The variables x_i and u_i are defined as state variables and control variables, respectively.

A version of the cusp catastrophe theory, which is governed by one state variable and two control variables, was employed in this study, and the standard format of its potential function $V(x)$ is

$$V(x) = x^4 + ux^2 + vx, \tag{6.16}$$

where x is the state variable; and u and v are the control variables. As is shown in Fig. 6.6, a three-dimensional phase space can be established with these three variables as coordinate axes. The equilibrium surface is defined as

$$grad_x V = x^3 + ux + v = 0, \tag{6.17}$$

where $grad_x$ is the gradient of the state variable x. It is assumed that a point Q moves on the above sheet of the equilibrium surface in a controlled route, the projection of which onto the control space $u - v$ is smooth. However, when point Q reaches a fold in the equilibrium surface, it will jump directly to below the sheet instead of to the middle of the sheet, leading to a sudden change in the variable x and a singular point on the control space $u - v$. This process shows how discontinuous states appear in a system controlled by a smooth potential function. In addition, the singularity set can be found easily by taking the gradient on both sides of the equal sign in Eq. (6.17):

$$grad_x(grad_x V) = 3x^2 + u = 0. \tag{6.18}$$

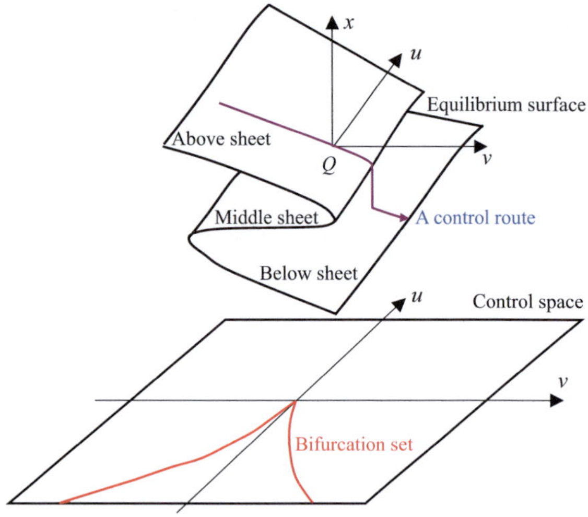

Fig. 6.6 Cusp catastrophic model

Finally, the bifurcation set can be obtained by projecting the singularity set onto the control space $u - v$ and eliminating the state variable x, which can be written as

$$8u^3 + 27v^3 = 0. \tag{6.19}$$

This is the set of equilibrium values of (x, u, v) for the cusp catastrophe theory, which brings about a sudden change in x.

6.3.2.2 A Mechanical Model of the Rock-Test Machine System

As is shown in Fig. 6.7, in order to study the sudden drop in the axial stress at the end of the cyclic disturbance phase, a simple mechanical model of rock-test machine system is proposed to employ the abovementioned cusp catastrophic theory. Equivalent springs are substituted for the top and bottom steel platens of the test machine and the rock sample. It is assumed that the load applied to the rock sample by the testing machine is P; the load bearing capacity of the rock sample is R; the deformation of the top and bottom steel platens of the test machine is a; and the deformation of the rock sample is u'. In the experiment in this study, the loading rate of the cyclic disturbance phase was smaller than that of a dynamic impact or blast, so the loading process can still be viewed as being quasi-static. Thus, the system is in equilibrium, that is,

$$P = R. \tag{6.20}$$

The load imposed on the spring in the test machine, which is within the elastic deformation range, can be written as

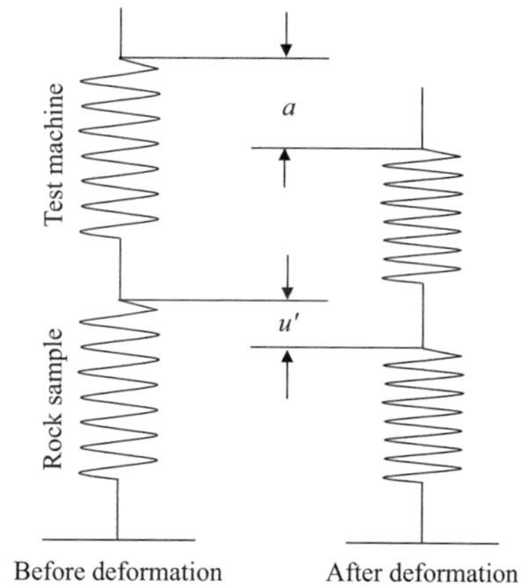

Fig. 6.7 Mechanical model of the rock-test machine system

$$R = ka, \tag{6.21}$$

where k is the stiffness of the loading platens.

In terms of the rock sample spring, its force–deformation relationship is notably non-linear. Tang (1993) proposed a force–deformation relationship for a sandstone sample:

$$R = \frac{EA_1}{L} u' \exp\left(-\frac{u'}{u_0}\right), \tag{6.22}$$

where E is the elasticity modulus; u_0 is the value of the deformation corresponding to the maximum loading capacity; and A_1 and L are the cross-sectional area and height of the sandstone sample, respectively.

Thus, the potential function of the rock-test machine system can be expressed in terms of its total energy:

$$V = \int_0^{u'} \frac{EA_1}{L} u' \exp\left(-\frac{u'}{u_0}\right) du' + \frac{1}{2} k\left(a - u'\right)^2. \tag{6.23}$$

On the right-hand side of Eq. (6.23), the first term and the second term are the sum of the energy absorbed by the rock sample and the sum of the energy absorbed by the test machine during the cyclic disturbance, respectively.

By proposing a similar mechanics model of the rock-rock (hypocenter-surrounding) system, Wang et al. (2006) concluded that the necessary unstable condition is completely determined by the internal characteristics of the system itself, and they deduced a failure criterion for the instability of their system, which can be introduced for the rock-test machine system used in this study. Their failure criterion has the following form:

$$k - \lambda \leq 0, \tag{6.24}$$

where λ is the instantaneous stiffness of the rock sample, which corresponds to the slope of the stress–strain curve at the stress drop point. In this study, this inequality indicates that the failure criterion of the cyclic disturbance test is that the stiffness of the top and bottom steel platens k is less than the instantaneous stiffness of the rock sample λ.

6.3.2.3 Analysis of the Sudden Jump in Stress at the End of the Cyclic Disturbance Phase

Based on the cusp catastrophe theory and the mechanical model of the rock-test machine system described above, to analytically obtain a bifurcation set for the sudden jump in stress at the end of the cyclic disturbance phase, we only need to derive an equation of the same form as Eq. (6.17) and determine the parameters in it.

Equation (6.23) can be transformed into the standard format for a potential function by expanding the power series and by omitting all of the items above the fourth power. Thus, when the rock sample becomes unstable during cyclic disturbance, the unbalanced load acting on the rock sample F can be expressed as

$$F = -\frac{\partial V}{\partial u'} = -c\left(x^3 + ux + v\right), \tag{6.25}$$

where $c = 4\lambda u_0^2/3$. If the state variable is taken as $x = (u' - u_1)/u_1$, where u_1 is the displacement of the stress drop point on the stress–strain curve, we can define $\dot{u}' = u_1\dot{x}$ and $\ddot{u}' = u_1\ddot{x}$, where $\dot{u}'$ and $\ddot{u}'$ are the velocity and acceleration of the rock sample during deformation, respectively.

In the cyclic disturbance phase, the total axial load applied to the rock sample can be denoted as $A_0 + A\sin\omega t$, where A_0 is the designated load corresponding to the set target point in the experiment, $A\sin\omega t$ is the cyclic disturbance load, t is time, and A and ω are the amplitude and frequency of the cyclic disturbance load, respectively. Then, the unbalanced load acting on the rock sample F at the end of the cyclic disturbance phase can be written as

$$F = m\ddot{u}' + \mu\dot{u}' - (A_0 + A\sin\omega t), \tag{6.26}$$

where m is the mass of the rock sample, and μ is the damping coefficient of the rock sample. By making the simplifying assumption that the damping load of the rock sample during the cyclic disturbance process is linearly related to the velocity, we obtain $\dot{u}' = u_1\dot{x}$. In terms of these parameters, the unbalanced load expression (Eq. 6.25) becomes

$$A_0 + A\sin\omega t = mu_1\ddot{x} + \mu u_1\dot{x} + c\left(x^3 + ux + v\right). \tag{6.27}$$

By defining $\eta = \mu/m$, $\alpha = c/m\mu$, $\omega_0^2 = cu/mu_1$, $d = cv/mu_1$, $e = A_0/mu_1$, and $f = A/mu_1$, Eq. (6.27) can be written as

$$\ddot{x} + \eta\dot{x} + \alpha x^3 + \omega_0^2 x + d = e + f\sin\omega t. \tag{6.28}$$

The deformation of the rock sample u' can be expressed as

$$u' = H_0 + H\sin(\omega t + \varphi), \tag{6.29}$$

where φ is the initial phase; and H_0 and H are the displacement before the cyclic disturbance and the amplitude of the displacement caused by the cyclic disturbance, respectively. Thus, the solution to Eq. (6.28) can be assumed to be

$$\begin{aligned} x &= \frac{u' - u_1}{u_1} = \frac{H_0 + H\sin(\omega t + \varphi) - u_1}{u_1} \\ &= \frac{H_0 - u_1}{u_1} + \frac{H}{u_1}\sin(\omega t + \varphi) = H_0' + H'\sin(\omega t + \varphi). \end{aligned} \tag{6.30}$$

In addition, the relationship between H_0 and A_0 can be obtained from Eq. (6.22) as

$$A_0 = \frac{EA_1}{L}H_0\exp\left(-\frac{H_0}{u_0}\right). \tag{6.31}$$

By substituting Eqs. (6.30) and (6.31) into Eq. (6.28), omitting the term with $\sin 3\omega t$ and the second term containing harmonics, and separately equalizing the constant term and the coefficients before the terms $\sin\omega t$ and $\cos\omega t$ on both sides of the equation, the following equations can be obtained:

$$\begin{cases} H_0'\omega_0^2 + \alpha H_0'^3 + d = e \\ H'\left[\left(\omega_0^2 - \omega^2 + 3\alpha H_0'^2\right) + \frac{3}{4}\alpha H'^2\right]\cos\varphi - \eta H'\omega\sin\varphi = f \\ H'\left[\left(\omega_0^2 - \omega^2 + 3\alpha H_0'^2\right) + \frac{3}{4}\alpha H'^2\right]\sin\varphi + \eta H'\omega\cos\varphi = 0 \end{cases}. \tag{6.32}$$

Among them, the first equation is the compatible equation between the internal structure of the rock sample and the initial conditions at the end of the quasi-static loading phase. It is not taken into consideration herein because it does not affect the analysis of the rock catastrophes. It can be derived from the second and third equations that

$$H'^2\left(\omega_0^2 - \omega^2 + 3\alpha H_0'^2 + \frac{3}{4}\alpha H'^2\right)^2 + \eta^2 H'^2\omega^2 = f^2. \tag{6.33}$$

By carrying out differential homeomorphism on Eq. (6.33) (Saunders 1980), we obtain

$$\chi^3 + \chi u + v = 0, \tag{6.34}$$

in which:

$$\chi = H'^2 + \frac{8\left(\omega_0^2 - \omega^2 + 3\alpha H_0'^2\right)}{9\alpha}, \tag{6.35}$$

$$u = \frac{16\left[3\eta^2\omega^2 - \left(\omega_0^2 - \omega^2 + 3\alpha H_0'^2\right)^2\right]}{27\alpha^2}, \tag{6.36}$$

and

$$v = \frac{-16\left\{\left[8\left(\omega_0^2 - \omega^2 + 3\alpha H_0'^2\right)^2 + 9\eta^2\omega^2\right] + 81\alpha f^2\right\}}{729\alpha^3}. \tag{6.37}$$

Therefore, Eq. (6.34) is the standard format of the equilibrium surface equation during the cyclic disturbance of the rock sample based on the cusp catastrophe theory, and the bifurcation set (Eq. 6.19) and the parameters in it can also be determined. Moreover, the cusp catastrophe theory may provide a potential theoretical explanation for the sudden instability or failure of a rock sample subjected to cyclic disturbance.

In short, the sudden instability of rock in the process of cyclic dynamic disturbance is due to the deterioration of its stiffness caused by the nucleation of microcracks. However, volumetric strain calculated by axial and radial strains at the end of the cyclic dynamic disturbance stage is positive, indicating that the volume expansion of rock sample does not occur at the end of the cyclic dynamic disturbance stage, that is, the rock is in a state of volume compression. This compaction state also improves the deformation resistance of rock samples to a certain extent. As a result, the volume compaction is superior to the microcracks nucleation, and ultimately leads to the improvement of the deformation resistance or instantaneous stiffness of the rock sample. According to the cusp catastrophe theory, the sudden instability phenomenon in the process of rock cyclic dynamic disturbance is caused by the fact that the stiffness of the testing machine is less than the instantaneous stiffness of the rock sample, which is mainly characterized by the displacement between the rock sample and the testing machine. If the cause of sudden instability of rock sample is extended to mechanical rock breaking in actual engineering, rock breaking tools continue to impose cyclic dynamic disturbance loads on rocks under high ground stress. In this process, the cracks in the tunnel surrounding rock continue to nucleate, the volume is gradually compressed to resist the cyclic dynamic disturbance loads imposed by the tool, and the instantaneous stiffness of the surrounding rock is finally increased. Once the stiffness of the rock breaking tool is less than the instantaneous stiffness of the surrounding rock, sudden instability occurs, such as rockburst.

6.4 Chapter Summary

In order to explain the mechanism of progressive damage and final failure of rock samples in the quasi-static phase and cyclic dynamic disturbance phase of rock samples in the triaxial cyclic disturbance experiments, this chapter firstly defined damage variables, and established a statistical damage constitutive equation of rock in quasi-static loading phase. Then, the mechanism of progressive damage in the stage of cyclic dynamic disturbance phase is given. Finally, the nucleation phenomenon of microcracks in the cyclic dynamic disturbance phase is analyzed, and the abrupt instability of the rock sample at the end of cyclic dynamic disturbance phase is explained using cusp catastrophe theory. The main conclusions are as follows:

1. According to the assumption that the strength of rock microbodies follows Weibull distribution and the failure conditions of the rock microbodies conform to the Hoek–Brown strength criterion, a damage evolution equation of the red sandstone sample in the quasi-static loading stage is established. The theoretical curve plotted by the constitutive equation have a good agreement with the experimental curves. The damage variable in the constitutive relationship shows an "S" shape with the evolution trend of axial strain, which accords with the characteristics of damage evolution of the rock samples under conventional triaxial compression.
2. The damage evolution process in the cyclic dynamic disturbance phase of the tests is the result of the combination of creep damage evolution under high stress and periodic cyclic dynamic disturbance. The mechanism of creep damage process at high stress level is that one or several dominant cracks in the rock expand at a relatively slow speed, and finally form a flake debris failure form. The damage mechanism of cyclic disturbance load is that the stiffness of rock material deteriorates compared with that without cyclic dynamic disturbance load.
3. Based on the cusp catastrophe theory and the mechanical model of the system of rock-test machine, a bifurcation set of the sudden jump of stress at the end of cyclic disturbance phase can be analytically obtained, which make a potential theoretical explanation for sudden instability or failure of rock subjected to cyclic disturbance.

References

Alejano LR, Bobet A (2012) Drucker-Prager criterion. Rock Mech Rock Eng 45:995–999. https://doi.org/10.1007/s00603-012-0278-2

Aliha MRM, Hosseinpour GR, Ayatollahi MR (2013) Application of cracked triangular specimen subjected to three-point bending for investigating fracture behavior of rock materials. Rock Mech Rock Eng 46(5):1023–1034. https://doi.org/10.1007/s00603-012-0325-z

Atkinson BK, Meredith PG (1987) The theory of subcritical crack growth with applications to minerals and rocks. Fract Mech Rock 111–166. https://doi.org/10.1016/B978-0-12-066266-1.50009-0

Cai M, Kaiser PK, Tasaka Y et al (2004) Generalized crack initiation and crack damage stress thresholds of brittle rock masses near underground excavations. Int J Rock Mech Min Sci 41(5):833–847. https://doi.org/10.1016/j.ijrmms.2004.02.001

Chai WG (2009) Experimental research on damage evolution and fail of granite under unloading confining pressure. China University of Mining and Technology, Beijing

Chang S-H, Lee C-I (2004) Estimation of cracking and damage mechanisms in rock under triaxial compression by moment tensor analysis of acoustic emission. Int J Rock Mech Min Sci 41(7):1069–1086. https://doi.org/10.1016/j.ijrmms.2004.04.006

Chen X, Yu J, Tang CA et al (2017) Experimental and numerical investigation of permeability evolution with damage of sandstone under triaxial compression. Rock Mech Rock Eng 50(6):1529–1549. https://doi.org/10.1007/s00603-017-1169-3

Chen S, Qiao C, Ye Q et al (2018) Comparative study on three-dimensional statistical damage constitutive modified model of rock based on power function and Weibull distribution. Environ Earth Sci 77(3):108. https://doi.org/10.1007/s12665-018-7297-6

Eberhardt E, Stead D, Stimpson B et al (1998) Identifying crack initiation and propagation thresholds in brittle rock. Can Geotech J 35(2):222–233. https://doi.org/10.1139/t97-091

Hoek E (1983) Strength of jointed rock masses. Géotechnique 33(33):187–223. https://doi.org/10.1680/geot.1983.33.3.187

Hoek E, Brown ET (1997) Practical estimates of rock mass strength. Int J Rock Mech Min Sci 34(8):1165–1186. https://doi.org/10.1016/S1365-1609(97)80069-X

Hoek E, Brown ET (2019) The Hoek–Brown failure criterion and GSI—2018 edition. J Rock Mech Geotech Eng 11(3):445–463. https://doi.org/10.1016/j.jrmge.2018.08.001

Joseph FL, Arno Z (2012) Mohr-Coulomb failure criterion. Rock Mech Rock Eng 45:975–979. https://doi.org/10.1007/s00603-012-0281-7

Kuksenko V, Tomilin N, Damaskinskaya E et al (1996) A two-stage model of fracture of rocks. Pure Appl Geophys 146(2):253–263. https://doi.org/10.1007/BF00876492

Lemaitre J (1985) A continuous damage mechanics model for ductile fracture. Trans ASME J Eng Mater Technol 107(107):83–89. https://doi.org/10.1115/1.3225775

Lemaitre J, Desmorat R (2005) Engineering damage mechanics. Springer, Berlin

Martin CD (1997) Seventeenth Canadian Geotechnical Colloquium: the effect of cohesion loss and stress path on brittle rock strength. Can Geotech J 34(5):698–725. https://doi.org/10.1139/cgj-34-5-698

Saunders PT (1980) An introduction to catastrophe theory. Cambridge University Press, Cambridge

Su CD, Xiong ZQ, Liu SW et al (2016) Experimental study of time-lag deformation and failure properties of coal under uniaxial compression. Rock Soil Mech 37(3):665–671. https://doi.org/10.16285/j.rsm.2016.03.007

Tang CA (1993) Catastrophe in rock unstable failure. China Coal Industry Publishing House, Beijing

Tang CA (1997) Numerical simulation of progressive rock failure and associated seismicity. Int J Rock Mech Min Sci 34(2):249–261. https://doi.org/10.1016/S0148-9062(96)00039-3

Tang CA, Liu H, Lee PKK et al (2000a) Numerical studies of the influence of microstructure on rock failure in uniaxial compression—Part I: Effect of heterogeneity. Int J Rock Mech Min Sci 37(4):555–569. https://doi.org/10.1016/S1365-1609(99)00121-5

Tang CA, Tham LG, Lee PKK et al (2000b) Numerical studies of the influence of microstructure on rock failure in uniaxial compression—Part II: Constraint, slenderness and size effect. Int J Rock Mech Min Sci 37(4):571–583. https://doi.org/10.1016/S1365-1609(99)00122-7

Wang SY, Lam KC, Au SK et al (2006) Analytical and numerical study on the pillar rockbursts mechanism. Rock Mech Rock Eng 39(5):445–467. https://doi.org/10.1007/s00603-005-0075-2

Xiao JQ, Ding DX, Jiang FL et al (2010) Fatigue damage variable and evolution of rock subjected to cyclic loading. Int J Rock Mech Min Sci 47:461–468. https://doi.org/10.1016/j.ijrmms.2009.11.003

Yang YS, Zhou H, Zhang CQ et al (2011) Experimental investigation on time-lag failure properties of marble under uniaxial compressive test. Rock Soil Mech 32(9):2714–2720. https://doi.org/10.3969/j.issn.1000-7598.2011.09.026

Yin TB, Lv B, Li X et al (2018) Effect of thermal treatment on the mode I fracture toughness of granite under dynamic and static coupling load. Eng Fract Mech 199:143–158. https://doi.org/10.1016/j.engfracmech.2018.05.035

Yu MH (2004) Unified strength theory and its applications. Springer, Berlin

Yu MH, Zan YW, Zhao J et al (2002) A unified strength criterion for rock material. Int J Rock Mech Min Sci 39(8):975–989. https://doi.org/10.1016/S1365-1609(02)00097-7

Yu MH, Kolupaev VA, Li YM et al (2011) Advances in unified strength theory and its generalization. Procedia Eng 10:2508–2513. https://doi.org/10.1016/j.proeng.2011.04.413

Zheng YR (2002) Plastic mechanics of rock and soil. China Architecture & Building Press, China

Zheng Q, Qian J, Zhang CS (2024) Velocity tomography of cross-sectional damage evolution along rock longitudinal direction under uniaxial loading. Tunn Undergr Space Technol 143:105503. https://doi.org/10.1016/j.tust.2023.105503

Zhou GL (1999) Complete stress-strain behavior for shear failure of rocks. University of Hong Kong. https://doi.org/10.5353/TH_B4257567

Chapter 7
Numerical Model of Progressive Damage and Failure of Rock Triggered by Cyclic Dynamic Disturbance

7.1 Introduction

As mentioned in the previous chapters, cyclic dynamic disturbance can induce damage and failure of rocks under high stress, but the use of numerical simulation to reproduce this process is still one of the current research difficulties. The cyclic dynamic disturbance experiments are often expensive and time-consuming. The theoretical models always show a departure from experimental observation because of the oversimplification of the rock breaking mechanism. Hence, numerical techniques based on continuum methods such as the finite element method (FEM), the discontinuous method such as discrete element method (DEM), and the hybrid method such as FEM-DEM, which are economical and convenient alternatives, offer new avenues to mimic the failure process of rock materials subjected to cyclic loading–unloading. A summary of different numerical methods applied to modeling rock materials' behavior under cyclic loading by some researchers is provided in Table 7.1. Fu et al. (2020) employed a FEM model for reproducing the cyclic loading experiment. They affirmed that the strengthening effect of rock subjected to cyclic loading could be reliably replicated using FEM with a cohesion variation and frictional strengthening (CVFS) model. Xi et al. (2021) developed a numerical method to model rock fracture induced by hydraulic pulses considering rock fatigue using in-house FORTRAN scripts and the ABAQUS solver. Still, the model cannot simulate the generation of a complex fracture network. Ma et al. (2017) provided a conceptual model describing the general shear behavior of rock joints under cyclic loading and the Barton–Bandis joint model considering the surface roughness degradation in 2D DDA (discontinuous deformation analysis). Dai et al. (2018) used a block-based DEM software—the Universal Distinct Element Code (UDEC), to replicate the stress–strain behavior and micro- and macro-mechanical damage evolution of rocks in loading and unloading conditions. The UDEC model can model the intact rock blocks and the interactions that occur along joints, but it cannot be used for predicting the fatigue life of

X. Chen, *Damage and Failure of Rock Triggered by Dynamic Disturbance*,
https://doi.org/10.1007/978-981-96-9486-0_7

rock. Song et al. (2019a, b) established discrete element models based on a particle-based DEM software PFC (particle flow code) to investigate the breaking process of sandstone under cyclic loading. The quantitative consistency between experimental and numerical results indicates that the proposed constitutive can reflect sandstone's damage evolution under cyclic loading. Liu et al. (2021) presented a combined finite-discrete element (FEM-DEM) procedure for investigating the fatigue behavior of jointed soft rock samples, and a suite of rock failure simulations at varying stress ratios and joint dip angles verified that the dynamic properties under cyclic loading can be well captured using FEM-DEM method. However, proven numerical modeling of the progressive damage and failure of rock submitted to cyclic dynamic disturbance is relatively scarce, especially considering the strength degradation of mesoscopic finite elements.

Therefore, this chapter attempts to introduce a numerical model to investigate the damage and collapse of rock under cyclic dynamic disturbance. A novel assumption that the strength of each mesoscopic finite element degrades with cyclic numbers as a linear function and implant this assumption under the framework of RFPA2D is proposed. Afterward, the reliability of the proposed model is examined by numerically performing cyclic loading tests for sandstone samples. Then this model is employed to investigate the damage progressivity of underground tunnels induced by the remote vertical cyclic loading.

7.2 Novel Model for Simulating Progressive Damage of Rock Subjected to Cyclic Loading Under the Framework of the RFPA2D Code

7.2.1 Theory of the RFPA2D Code

7.2.1.1 Heterogeneity Consideration

From 1997 onward, the RFPA2D code has gained international recognition for simulating rock materials' damage and failure process under various boundary conditions (Tang 1997; Yang et al. 2004; Liang et al. 2012; Tang et al. 2020). This code employs a finite element method solver with four-node isoparametric elements in meshes to calculate and analyze stress, strain, and displacement fields. To take the heterogeneity of rock materials on the mesoscopic scale into consideration, the mechanical properties of the material (e.g., the elastic modulus, Poisson's ratio, and uniaxial compressive strength) assigned to four-node elements of the numerical specimen are assumed to conform to a Weibull distribution, of which the probability density function is defined by

$$\varphi(\alpha)=\frac{m}{\alpha_0}\cdot\left(\frac{\alpha}{\alpha_0}\right)^{m-1}\cdot\exp\left[-\left(\frac{\alpha}{\alpha_0}\right)^{m}\right] \tag{7.1}$$

Table 7.1 Summary of different numerical methods applied on modeling rock materials behavior under cyclic loading

Numerical method	Model	Advantage	Disadvantage	References
FEM (RFPA2D)	Cohesion variation and frictional strengthening (CVFS) model	Be suitable for modeling strengthening effect during cyclic loading	Narrow range of applications	Fu et al. (2020)
FEM (ABAQUS)	Coupling of rock *S-N* curves with the fluid cohesive crack model	Have good performance in the simulation of hydraulic cyclic loading	• Cannot simulate the generation of a complex fracture network • Be short of considering fracture and multiple cracks	Xi et al. (2021)
DDA	The Barton–Bandis joint model together with the conceptual model describing the general shear behavior of rock joints	Be capable of predicting the varying shear stress and dilation behaviors of rock joints under cyclic shear loading	Cannot be used for simulating mechanical behaviors under cyclic compressive loading	Ma et al. (2017)
DEM (UDEC)	Model based on elastic and unbreakable voronoi blocks	Have capacity to model the intact rock blocks as well as the interactions that occur along joint	Cannot be used for predicting fatigue life of rock	Dai et al. (2018)
DEM (PFC)	Nonlinear parallel-bonded stress corrosion (NPSC) model	Good at solving the large deformation motion of intermittent systems	Cannot consider the deformation of particles or blocks	Song et al. (2019a, b)
FEM-DEM		Be appropriate for modeling the continuous to discontinuous failure process of rocks	Need to artificially define crack element and uncrack element on the numerical model before loading	Liu et al. (2021)

where α is a given mechanical parameter of rock material; α_0 is the scale parameter of the Weibull distribution, representing the mean value of the given mechanical parameter; *m* is the shape parameter of the Weibull distribution that controls the distribution shape of the given mechanical parameter. As shown in Fig. 7.1, the functional curves have different distribution shapes due to different values of m in the Weibull distribution. In the RFPA2D code, m is defined as the degree of rock heterogeneity at the meso-level. A larger value of *m* ($m > 1$) means that the given mechanical parameters of the mesoscopic elements are more concentrated in a small

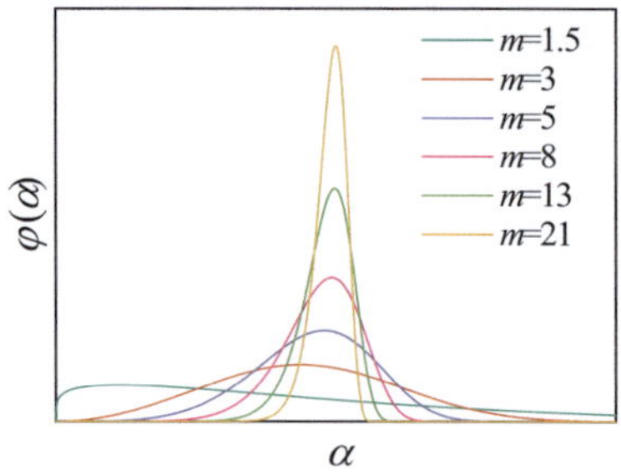

Fig. 7.1 Probability density functional images of the Weibull distribution with different shape parameters m

range around their mean value α_0 and the interior of the rock material is relatively more homogeneous, and vice versa.

7.2.1.2 Statistical Damage Constitutive Model

According to the principle of strain equivalence in damage mechanics (Lemaitre and Desmorat 2005), the constitutive relationship of a one-dimensional linear elastic damage material is expressed as

$$\sigma = E\varepsilon = E_0\varepsilon(1 - D), \tag{7.2}$$

where σ and ε are the stress and strain of the damaged material under a one-dimensional load, respectively, E is the elastic modulus of the damaged material, E_0 is the equivalent elastic modulus of the material in an entire state, and D is the damage variable. Therefore, the elastic modulus E is applicable for describing the damage evolution of the damaged material as follows:

$$E = E_0(1 - D). \tag{7.3}$$

The RFPA2D assumes that all mesoscopic elements that constitute the numerical rock model are linearly elastic, isotropic, and undamaged before uniaxial or multiaxial loading. Figure 7.2 shows the elastic constitutive model of a mesoscopic element under uniaxial loading (tensile or compressive) conditions. According to sign convention in rock mechanics, it is usually stipulated that compression is in the positive direction and tension is in the negative direction. It can be seen from Fig. 7.2 that on the tension side in the third quadrant of the model, the mesoscopic element is linearly elastically deformed and not yet damaged in the tensile strain range from point O to ε_{t0}. According to the maximum tensile strain criterion, when the tensile stress reaches f_t, the mesoscopic element begins to damage, and the axial stress decreases steeply, then the strain softens into constant residual stress f_{tr} up to the ultimate tensile strain ε_{tu}. Therefore, in the state of tensile stress, the damage variable D of the element can be expressed as

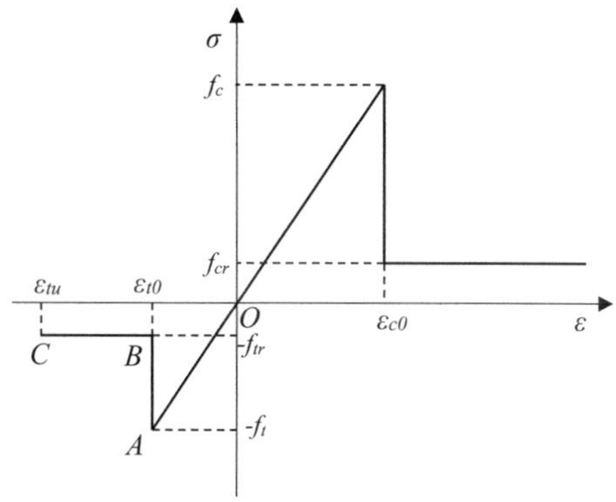

Fig. 7.2 Mechanical behavior model of elements under uniaxial loading in RFPA (Tang et al. 2002)

$$D = \begin{cases} 0 & \varepsilon_{t0} \le \varepsilon \\ 1 - \frac{f_{tr}}{E_0 \varepsilon} & \varepsilon_{tu} \le \varepsilon \le \varepsilon_0 \\ 1 & \varepsilon \le \varepsilon_{tu} \end{cases} \tag{7.4}$$

Likewise, on the compression side in the first quadrant of the model, the damage variable D can be written as

$$D = \begin{cases} 0 & \varepsilon \le \varepsilon_{c0} \\ 1 - \frac{f_{cr}}{E_0 \varepsilon} & \varepsilon_{c0} \le \varepsilon \end{cases}, \tag{7.5}$$

where f_{cr} is compressive residual stress of the mesoscopic element and ε_{c0} is the maximum principal strain indicating that damage to the element begins to occur at this value.

The Mohr–Coulomb criterion is used to express the second damage threshold in RFPA2D code when the mesoscopic element is under shear stress or multi-axial loading conditions, which can be described as

$$\sigma_1 - \sigma_3 \frac{1 + \sin\varphi}{1 - \sin\varphi} \ge f_c, \tag{7.6}$$

where σ_1 and σ_3 are the maximum and minimum principal stresses, respectively; φ is the internal friction angle of the mesoscopic element, and f_c is the uniaxial compressive strength of the mesoscopic element. Substituting Eq. (7.6) into the generalized Hooke law, the maximum principal strain ε_{c0} can be written as

$$\varepsilon_{c0} = \frac{1}{E_0}\left[f_c + \frac{1 + \sin\varphi}{1 - \sin\varphi}\sigma_3 - \nu(\sigma_2 + \sigma_3) \right], \tag{7.7}$$

where σ_2 and ν are intermediate principal stress and Poisson's ratio, respectively.

If it is assumed that the damage evolution of each mesoscopic element under three-dimensional loading conditions is only governed by the maximum compression principal strain ε_1, Eq. (7.5) can be further extended to describe damage evolution under triaxial stress substituting Eq. (7.6) into it. Thus, the extended formulation can be used to solve the shear damage problem under triaxial stress states, which is given by

$$D = \begin{cases} 0 & \varepsilon_1 \leq \varepsilon_{c0} \\ 1 - \frac{f_{cr}}{E_0 \varepsilon_1} & \varepsilon_{c0} \leq \varepsilon_1 \end{cases}. \tag{7.8}$$

7.2.2 *Novel Assumption of Strength Degradation of Mesoscopic Finite Elements*

From the abovementioned damage constitutive model of the widely used RFPA2D software, damage to the mesoscopic elements will not occur before the applied stress exceeds their certain threshold conditions—Eqs. (7.4), (7.5), and (7.8). For continuous cyclic loading in the same stress interval range, damage to the mesoscopic elements only occurs in the first cycle. No new damage occurring in the subsequent cycles unless the reload stress exceeds the previously applied stress (Bao et al. 2011). Thus, the damage progressivity during cyclic loading cannot be effectively numerically simulated with the existing RFPA2D code. Therefore, it is necessary to develop the code into an innovative numerical approach for modeling rock's progressive damage and failure under continuous cyclic loading.

To achieve this, a new assumption that the strength of each mesoscopic finite element degrades with cyclic numbers as a linear function is proposed by referring to the simple Palmgren–Miner linear rule (Lemaitre and Desmorat 2005). Figure 7.3 presents the schematic plot of the new assumption. Monotonic loading is first imposed on the numerical model in the calculation process. Each mesoscopic element will be calculated iteratively for t loading steps until the axial stress reaches $(\sigma_{max} + \sigma_{min})/2$, corresponding to point O to point A in the figure. Then, a cyclic loading–unloading mode is entered (from point A to point C′). In the cyclic loading phase, it is assumed that the strength of each mesoscopic element of the rock will continue to degrade according to a specific rule. In fatigue mechanics, there is a mathematical relationship between the fatigue life N_f and the absolute value of the difference between the cyclic loading and unloading upper limit stress σ_{max} and the lower limit stress σ_{min}, which is given by (Broek 1988)

$$N_f = C(|\sigma_{\max} - \sigma_{\min}|)^{\delta} \tag{7.9}$$

where C and δ represent the y-axis-intercept and the straight-line slope of the fatigue rate curve—region II, respectively (Mlikota et al. 2018).

Introducing the strength deterioration curve of the mesoscopic elements into Eq. (7.9), it can be furtherly expressed as a mathematical relationship between the

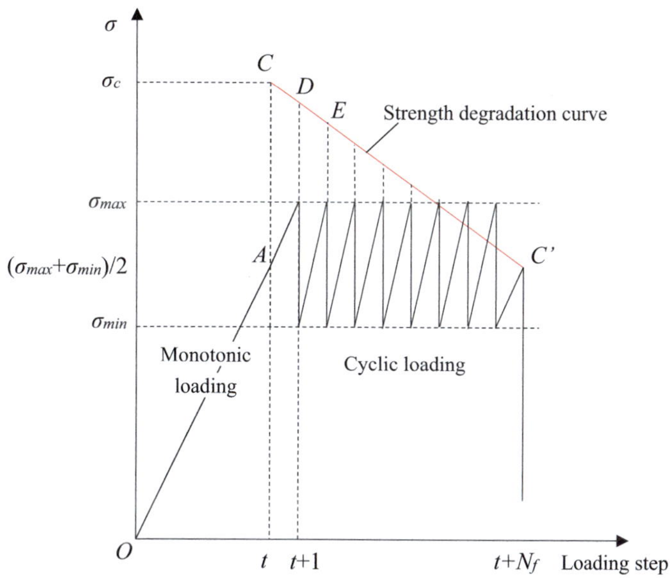

Fig. 7.3 Loading and unloading principle of mesoscopic elements in RFPA2D-Cycle code

number of loading–unloading cycles and the amplitude of the periodic load. It is assumed that the strength deterioration curve is linear (from point C to point C′), and the strength of each element is related to the stress level $(\sigma_{max} + \sigma_{min})/2$. At the starting point A, since the elements have not undergone cyclic loading and unloading, their uniaxial compressive strength can reach the undegraded uniaxial compressive strength σ_c; that is, the coordinate values of point C in Fig. 7.3 are (t, σ_c). At the end of cyclic loading and unloading C′, it is considered that the ultimate failure occurs at the strength of the stress level $(\sigma_{max} + \sigma_{min})/2$, meaning that the coordinate values of point C′ are $(t + N_f, (\sigma_{max} + \sigma_{min})/2)$. Thus, the equation for the strength degradation line CC′ can be obtained as

$$\sigma_{\text{cycle}} = \frac{\frac{\sigma_{\max}+\sigma_{\min}}{2} - \sigma_c}{N_f} \cdot (n - t) + \sigma_c \tag{7.10}$$

where σ_{cycle} is the uniaxial compressive strength of the element after n loading steps.

Substituting Eq. (7.9) into Eq. (7.10), the calculation principle formula of the uniaxial compressive strength of the mesoscopic element continuously deteriorates as the loading step increases is

$$\sigma_{\text{cycle}} = \frac{\frac{\sigma_{\max}+\sigma_{\min}}{2} - \sigma_c}{C(|\sigma_{\max} - \sigma_{\min}|)^{\alpha}} \cdot (n - t) + \sigma_c. \tag{7.11}$$

7.2.3 Flow Chart of the Developed Numerical Model

The numerical model approach based on the novel assumption that considers strength degradation of mesoscopic finite elements is implemented in the previous version of the RFPA2D code. The flow chart of numerical procedures is illustrated in Fig. 7.4. According to the loading and unloading rules of a mesoscopic element in the developed model, the computing process is mainly divided into two sections:

The first section analyzes the numerical model's stress, strain, deformation, and damage under the monotonic loading condition. After establishing the numerical calculation model, assigning materials parameters, and specifying boundary conditions for monotonic loading, element stiffness matrices and mass matrices can be built. With these matrices, linear elastic finite element equations can be iteratively solved to obtain data on the stress and strain of each element. Then, by employing the maximum tensile strain criterion and Mohr–Coulomb criterion, the combination of the principal stresses of each element is compared with their strength thresholds to examine whether the elements are irreversibly damaged or not. If irreversible damage to some elements occurs, those matrices will be rebuilt after damage process analysis, and the above procedures will be repeated on the model, which associates new parameters. Until no new damage occurs on any elements, the code will further check whether all monotonic loading steps are completed. If there are still some loading steps left, the boundary conditions for the next step will be modified, and the analysis procedures of monotonic loading are repeatedly executed in a loop.

The second section is an analysis of cyclic loading and unloading. Although the calculation process in this section closely parallels that of the first one, there are some differences between them. Before the second section executes, some cyclic loading and unloading parameters require input into the developed model, such as stress level, upper limit stress, lower limit stress, and the number of equivalent cycles per calculation step. Also, the uniaxial compressive strength of each element needs to be degraded according to Eq. (7.11) at the end of the current calculation step before entering the next calculation step in the loop computation.

7.3 Verification and Validation of the Modeling Approach

The reliability of the modeling approach for simulating progressive damage and failure of rock subjected to cyclic loading requires verification. Therefore, we set forth to numerically replicate the results of fatigue failure tests on red sandstone specimens in Chap. 4. According to the experimental method, red sandstone specimens were placed in a modified rock mechanical triaxial test system to withstand a specific amount of cyclic loading with different frequencies and different amplitudes. The waveform of the periodic loading was a sinusoidal wave. Three sets of data, listed in Table 7.2, were selected to conduct the numerical replication.

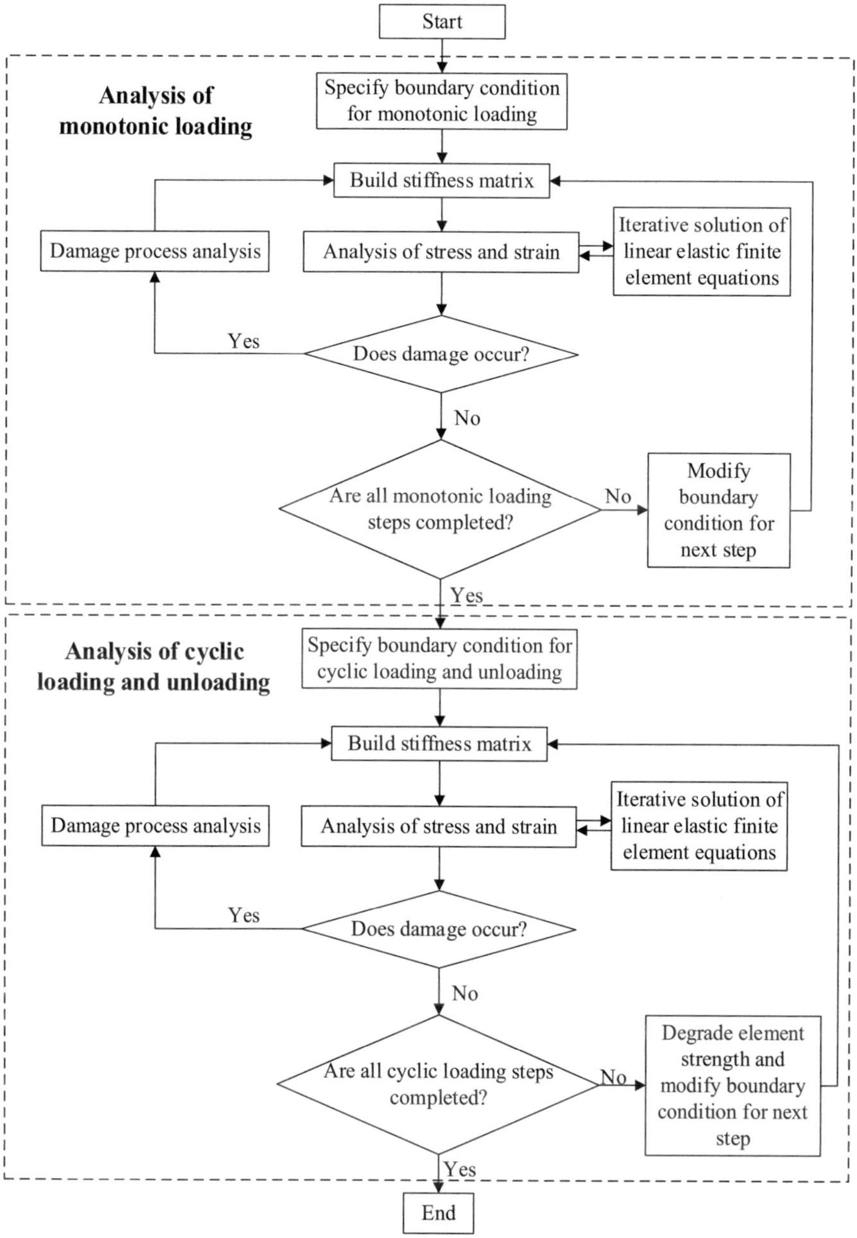

Fig. 7.4 Flow chart of numerical analysis in RFPA2D-Cycle code

Table 7.2 Test results of red sandstone samples under cyclic loading conditions

Sample no.	Frequency (Hz)	Amplitude (MPa)	Upper limit stress (MPa)	Lower limit stress (MPa)	Number of cycles
RS-10-1.0	10	1.0	56	54	16,720
RS-10-2.5	10	2.5	57.5	52.5	15,180
RS-10-5.0	10	5.0	60	50	3987

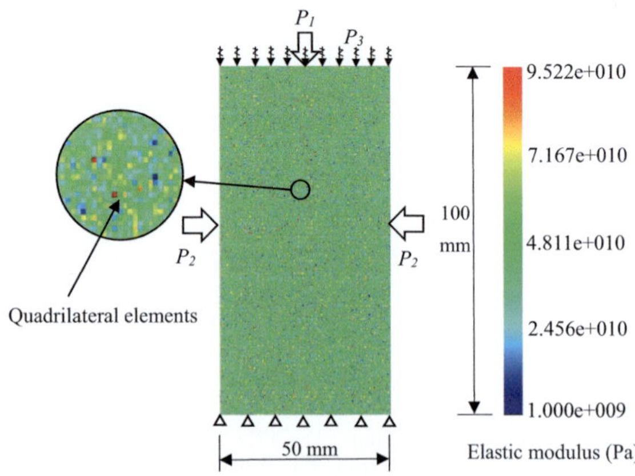

Fig. 7.5 Numerical model of a rock sample subjected to cyclic loading

7.3.1 Numerical Model Setup and Parameters

To numerically simulate the test conditions, the cylindrical specimen in the experiment was approximated by a two-dimensional axisymmetric (biaxial) model, i.e., a rectangular cross-section of dimensions 50 mm × 100 mm. As illustrated in Fig. 7.5, the model was discretized and meshed using $200 \times 400 = 80{,}000$ quadrilateral elements of the same size. P_1, P_2, and P_3 are axial loading, confining pressure, and cyclic loading. To correspond to the test conditions, P_1 was monotonically loaded to 55 MPa, which equals the stress level, and P_2 was set to 0 MPa, owing to the uniaxial compressive state. P_3 was input into the model based on the upper limit stress and lower limit stress values shown in Table 7.2. Some other material parameters of the numerical model are listed in Table 7.3.

7.3.2 Numerical Results

7.3.2.1 Damage Evolution Process

Figure 7.6 compares stress-time curves for the laboratory experiments from Chap. 4 with the stress-step curve simulated by the developed numerical model. It can be

Table 7.3 Parameters calibrated for the red sandstone specimen in the numerical model

Parameter	Value
Homogeneity index, m	3
Mean elastic modulus (GPa)	50
Poisson's ratio, v	0.25
Mean uniaxial compressive strength (MPa)	400
Friction angle (°)	30
Material parameter, C	23,259
Material parameter, δ	− 0.15549
Upper limit stress (MPa)	56; 57.5; 60
Lower limit stress (MPa)	54; 52.5; 50
Number of equivalent cycles per calculation step	70

seen from this figure that the curves simulated by the newly proposed model are in good accordance with the experimental curves. The verification results suggest that the newly developed computing model efficiently models stress-step curves of rock materials.

Figure 7.7 illustrates numerical results of the damage evolution process of sandstone samples under cyclic loading with three different amplitudes. Under the cyclic loading with three different amplitude levels at the same stress level, which is approximately 80% of the uniaxial compressive strength, i.e., $(\sigma_{max} + \sigma_{min})/2 = 55$ MPa, the elastic modulus of some elements in the numerical models gradually decreases (number of blue elements increases in Fig. 7.7). The damage slowly evolves until the numerical samples eventually break down to form a macroscopic rupture, making the model unable to continue calculations. Before damage occurs, some mesoscopic elements in the numerical rock samples show a downtrend in elastic modulus with the cyclic loading–unloading conditions continuously applied. These elements are distributed randomly throughout the plane of the numerical model. Then, some of those elements reach their damage thresholds and become damaged. The damaged elements further trigger some other elements nearby to become weak elements prone to damage. As a result, the damaged elements are concentrated, nucleation occurs, and macroscopic cracks are formed. The calculation steps of damage concentration occurring, as shown in Fig. 7.7a–c, are approximately at Step 252, Step 235, and Step 113, respectively. Investigations from Chap. 4 showed that the evolution of the damage variable, defined by the ratio of residual axial strain after n loading–unloading cycles to the ultimate residual axial strain at failure, increases sharply first and subsequently increases slowly. Xiao et al. (2009, 2010, 2011) also reported that the development of cracks passes through three phases—crack initiation, stable propagation, and unstable propagation—during cyclic loading–unloading. Accordingly, the damage evolution process numerically simulated using the new numerical model is in good agreement with their experimental results.

Regarding failure modes of the simulated results shown in Fig. 7.7, at the amplitude of 1.0 MPa, the failure occurs in the upper right corner of the numerical rock

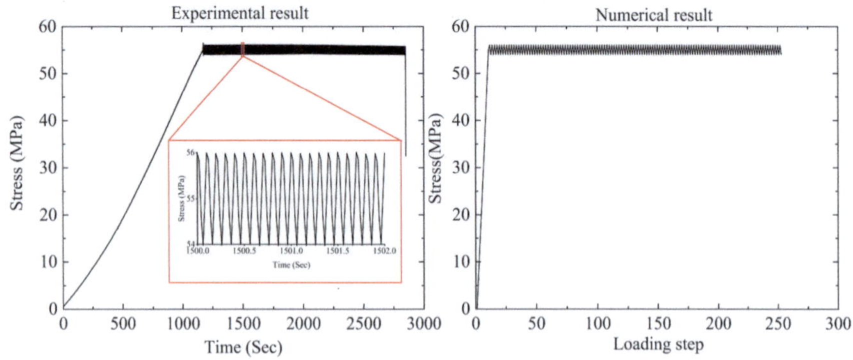

(a) Amplitude of 1.0 MPa

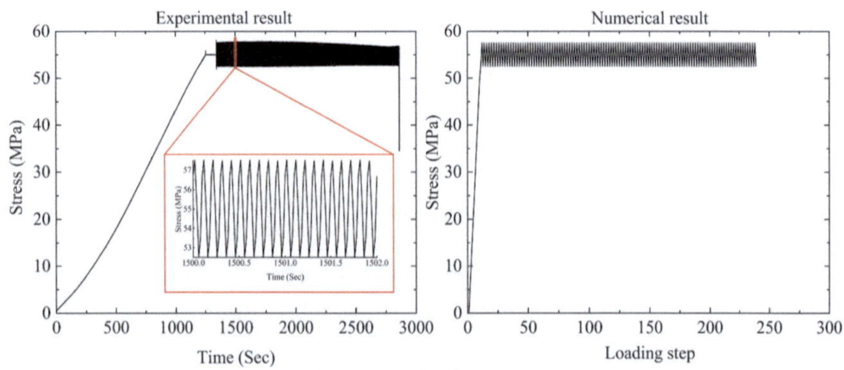

(b) Amplitude of 2.5 MPa

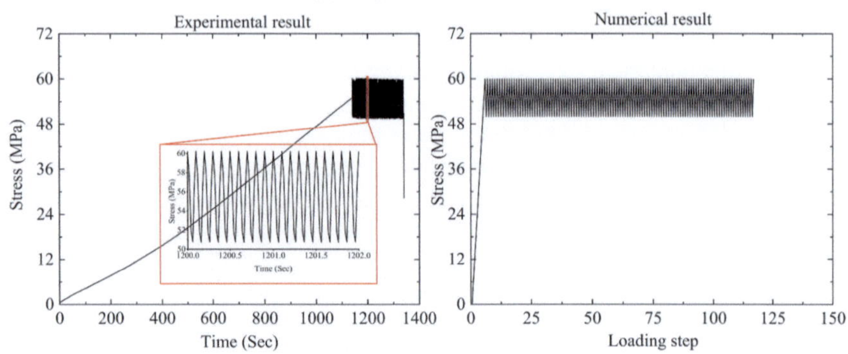

(c) Amplitude of 5.0 MPa

Fig. 7.6 Comparison of stress-time curves for the laboratory experiments with the RFPA2D-Cycle simulation

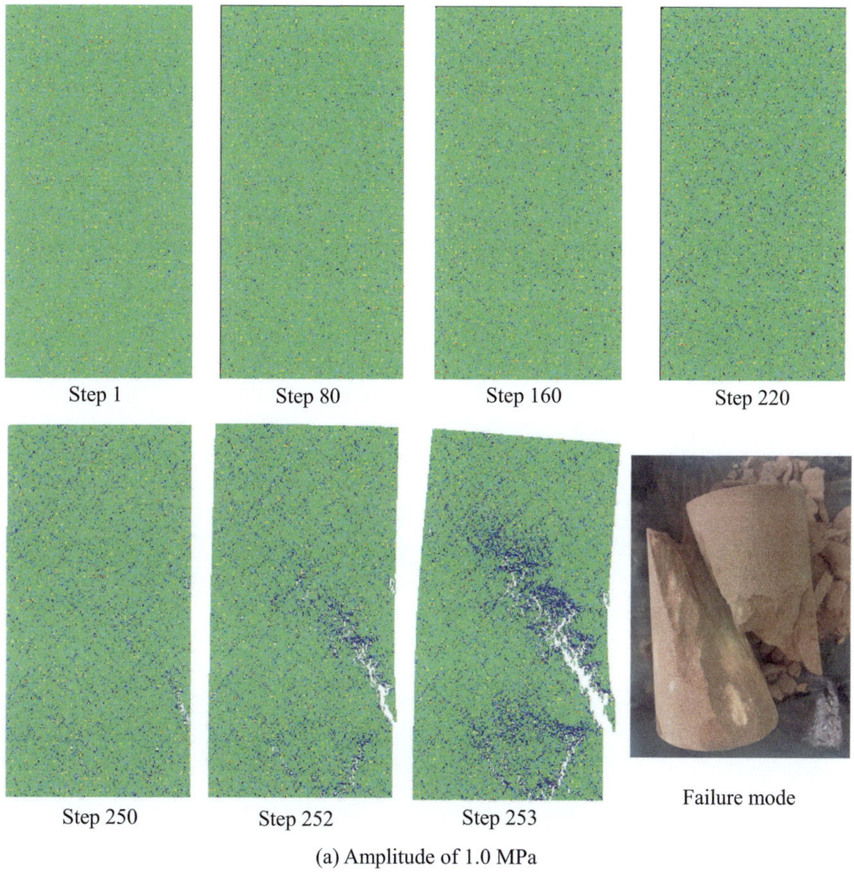

Fig. 7.7 Numerical results of damage evolution process of sandstone samples under cyclic loading

sample, leading to a local inclined fracture. At the amplitudes of 2.5 and 5.0 MPa, the numerical samples fail by modes of cataclasis. Therefore, to some extent, the simulated failure modes have similar characteristics to the failure modes obtained from the cyclic loading–unloading experiments in Chap. 4.

7.3.2.2 Analysis of AE Events

During deformation and damage in experiments or excavation, the released elastic energy of rock can be detected by some high accuracy sensors or transducers, called acoustic emission (AE) or microseismic monitoring techniques. Many studies have proven the effectiveness of using these techniques to determine the location of damage and observe damage evolution in rock mechanical experiments or rock engineering (Tang et al. 2010; Xiao et al. 2016; Feng et al. 2019). Based on this, it is

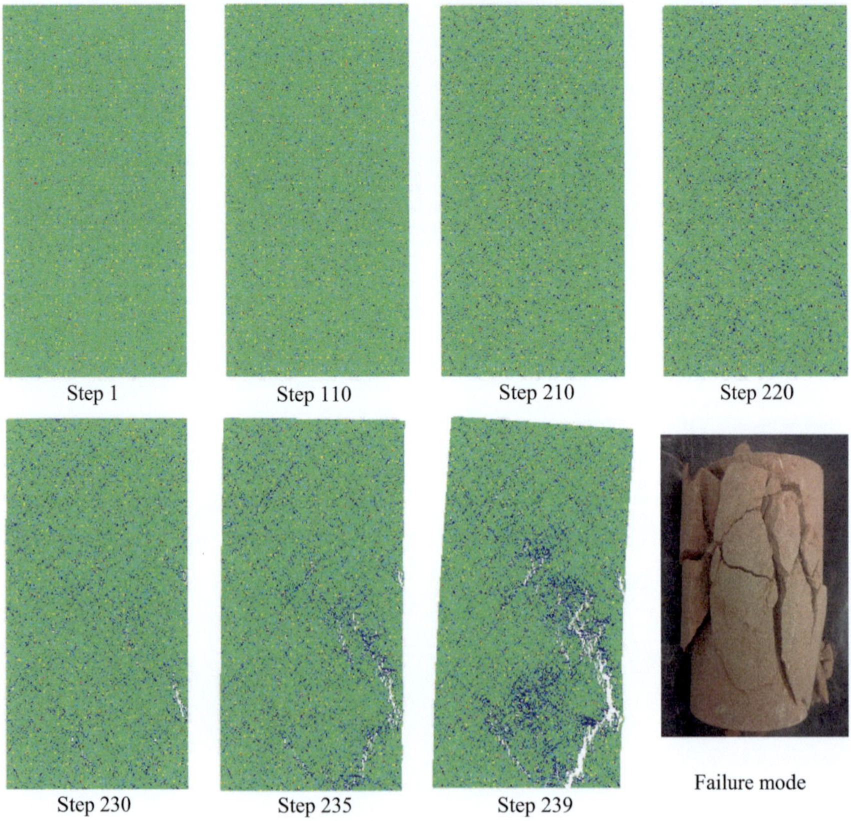

(b) Amplitude of 2.5 MPa

Fig. 7.7 (continued)

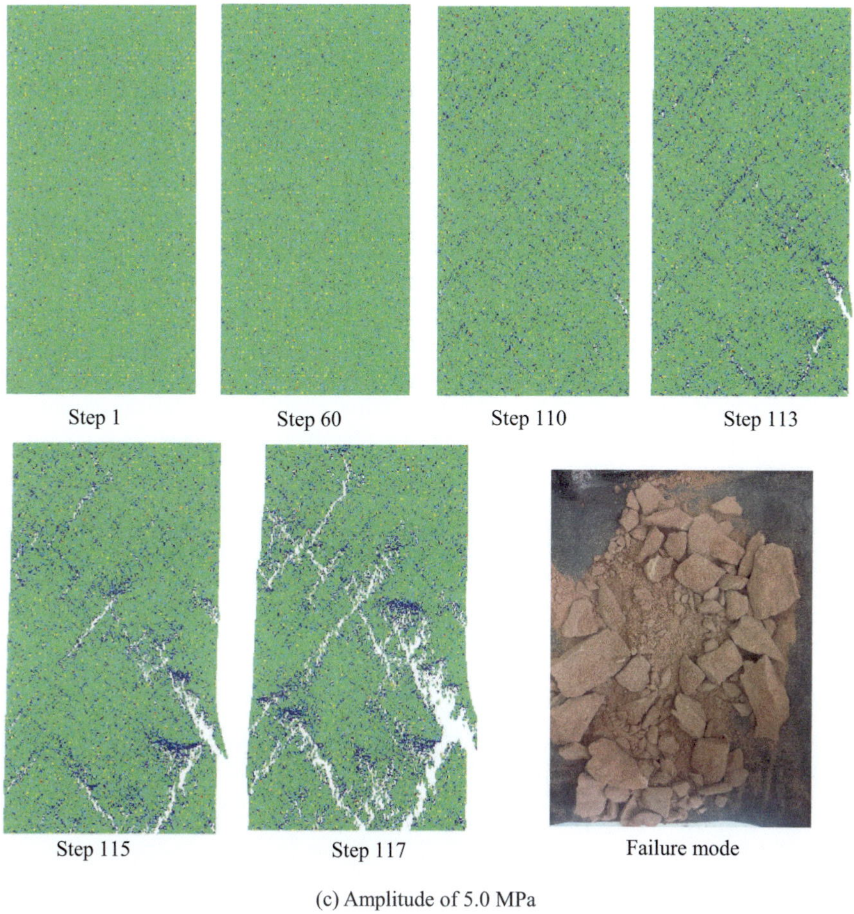

Fig. 7.7 (continued)

assumed that the number of seismic or AE events N_{AE} is directly proportional to the damage variable D, which is written by

$$N_{AE} \propto D \tag{7.12}$$

Accordingly, the AE events associated with the progressive failure can be numerically simulated by counting the number of damaged elements. The approximate magnitude of elastic energy release of those failure elements can also be evaluated using the element strengths.

Figure 7.8, which should be read in conjunction with Fig. 7.7, shows numerical results of the location of AE events in the numerical samples subjected to a continuous cyclic loading–unloading disturbance at the three different amplitude levels (1.0, 2.5, and 5.0 MPa). In Fig. 7.8, the white and red circles separately indicate the AE events triggered by shear compression damage and tensile damage of the mesoscopic elements at the current loading step. The centers of the circles mark the locations of the AE events, and the radiuses of the circles represent the relative magnitudes of the elastic energy release (EER). It is evident from Fig. 7.8 that, in the early stage, when the cyclic loads are applied, the AE events are randomly triggered on the numerical rock sample planes. The AE events caused by shear compression damage are more than those caused by tensile damage. As the cyclic loads continue to be applied, the AE events gradually increase, and the number of AE produced by shear compression damage in this stage is also more than that produced by tensile damage. The location distributions of these newly planted AE events are still relatively random without any apparent regularity. In the later stage of cyclic loading, the locations where the AE events are triggered start to concentrate on the local positions. The AE events are dominated by tensile damage of the mesoscopic elements, and the elastic energy released by tensile damage is much larger than that of shear damage.

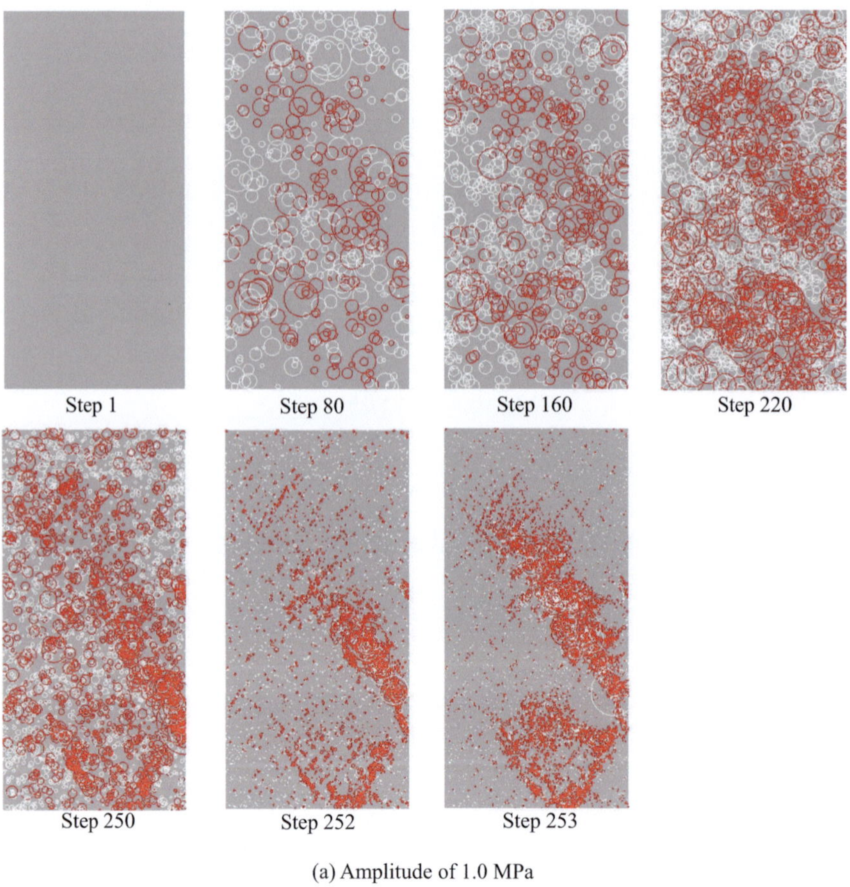

Fig. 7.8 Numerical results of location of AE events

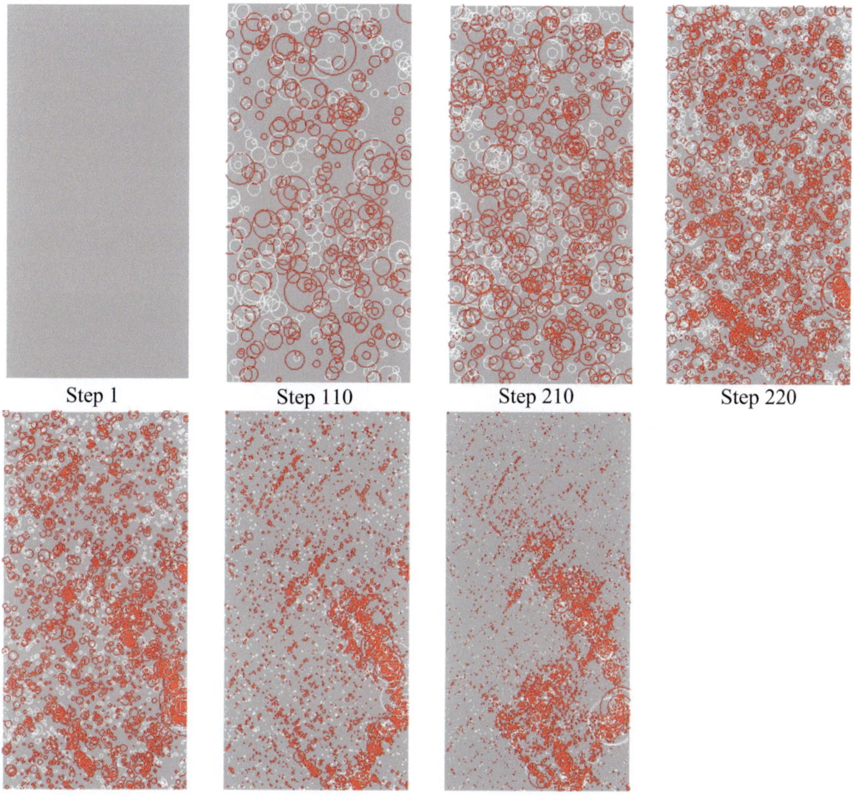

(b) Amplitude of 2.5 MPa

Fig. 7.8 (continued)

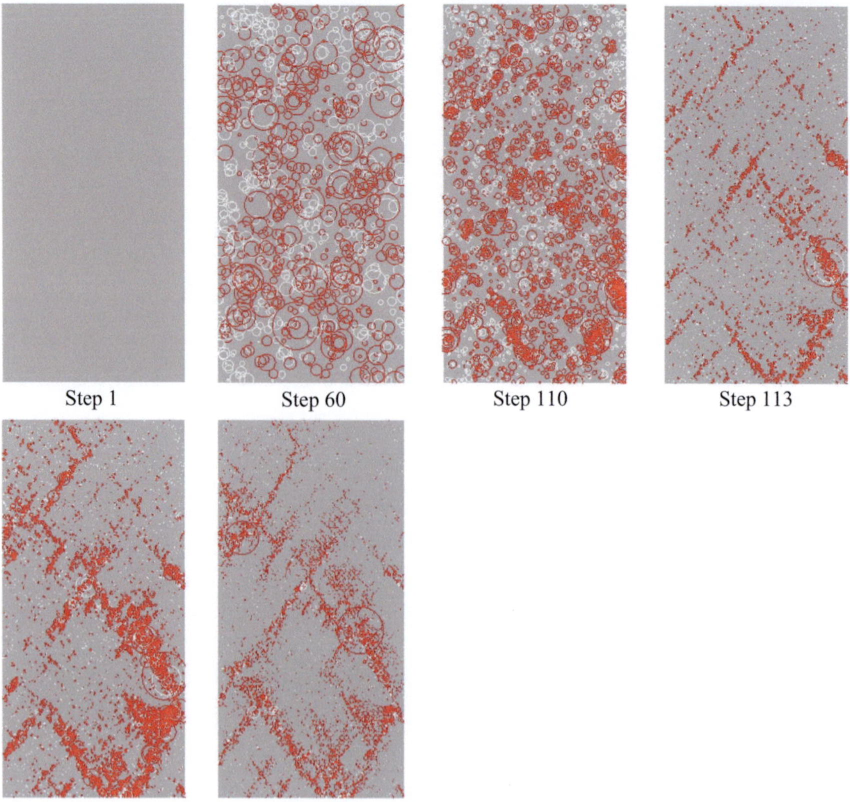

Step 1 Step 60 Step 110 Step 113

Step 115 Step 117

(c) Amplitude of 5.0 MPa

Fig. 7.8 (continued)

At the three different amplitude levels of 1.0, 2.5, and 5.0 MPa, the AE evolution inside the numerical rock samples is highly similar—from random generation across entire planes to concentrated generation at local locations. The distributions of the AE events are mostly clustered around the nucleation zone where the macroscopic cracks are formed. Hence, the numerically obtained results for the development of AE events shown in Fig. 7.8 follow the corresponding phenomenon as the damage evolution shown in Fig. 7.7.

Figure 7.9 illustrates numerical results of the stress, the AE counts, accumulated EER during progressive damage, and failure of rock samples subjected to cyclic loading with the three different amplitude levels (1.0, 2.5, and 5.0 MPa). In the experiment reported in Chap. 4, the loading process was divided into quasi-static and disturbance loading phases. The loading strategy in the numerical simulation follows the same experiment method; that is, the first ten loading steps are monotonically static loading, and the rest loading steps are cyclic loading. From Fig. 7.9a–c, excluding the first ten loading steps of static loading, a total of 243, 229, and 107 loading steps of cyclic loading with amplitudes of 1.0, 2.5, and 5.0 MPa are respectively imposed on the numerical rock samples before failure. Since each loading step is equivalent to 70 loading–unloading cycles in this numerical simulation, the number of cycles that the numerical rock samples have experienced are 17,010, 16,030, and 7490, respectively. In the experiment conducted in Chap. 4, the number of loading–unloading cycles is affected by amplitude and frequency. It decreases with the increasing amplitude at a specific frequency (shown in Table 7.1). Hence, the numerical simulation of the number of cyclic loads by the proposed numerical approach yields the same qualitative result as the experiment.

During cyclic loading in Fig. 7.9, AE events occur sporadically and intermittently in the early stage, and the numbers of AE counts in these loading steps are relatively small. However, in the last several loading steps before the failure of the numerical rock samples, the AE events occur continuously, and the number of occurrences of AE increases significantly, rising to several times or even dozens of times that of the previous loading steps. In addition, the accumulated EER of AE events shows an increasing trend in the whole cycle loading process. In the early stage, the accumulated EER increases slowly, but in the last few loading steps, the accumulated EER proliferates. It is indicated that the rock samples subjected to cyclic loading would produce many AE events and release a significant amount of elastic energy before failure and instability. In this sense, AE or microseismic monitoring techniques can be considered in actual engineering to monitor AE events produced or the elastic energy released during progressive damage of rock subjected to cyclic loading to achieve the purpose of early warning of failure or instability.

7.4 Discussion

7.4.1 Application of the Novel Numerical Model

Underground tunnels are subjected to horizontal and vertical ground stresses after excavating and stabilizing. Especially for some deep rock engineering, the ground stresses of the surrounding rock are relatively higher than that of the shallow tunnels. If a cyclic disturbance load, such as repeated traffic load on the ground, continuous low-frequency vibration load of heavy machinery, or cyclic disturbance of drilling machine is repeatedly applied in the vertical direction from the far-field under these high ground stress conditions, damage, and failure of the surrounding rock may be triggered, which often leads to large deformation, local instability, or even rockburst. In addition, vertical ground stress is generally not equal to horizontal ground stress in most in situ cases. This inequality causes the underground tunnels to be in a more complicated stress environment. To explore the instability process of underground tunnels induced by remote vertical cyclic loading, damage evolution around an underground tunnel for various lateral pressure coefficients of $k = P_h$ (MPa)/P_v (MPa), in

Fig. 7.9 Numerical results of stress, AE counts, accumulated elastic energy release (EER) versus loading step

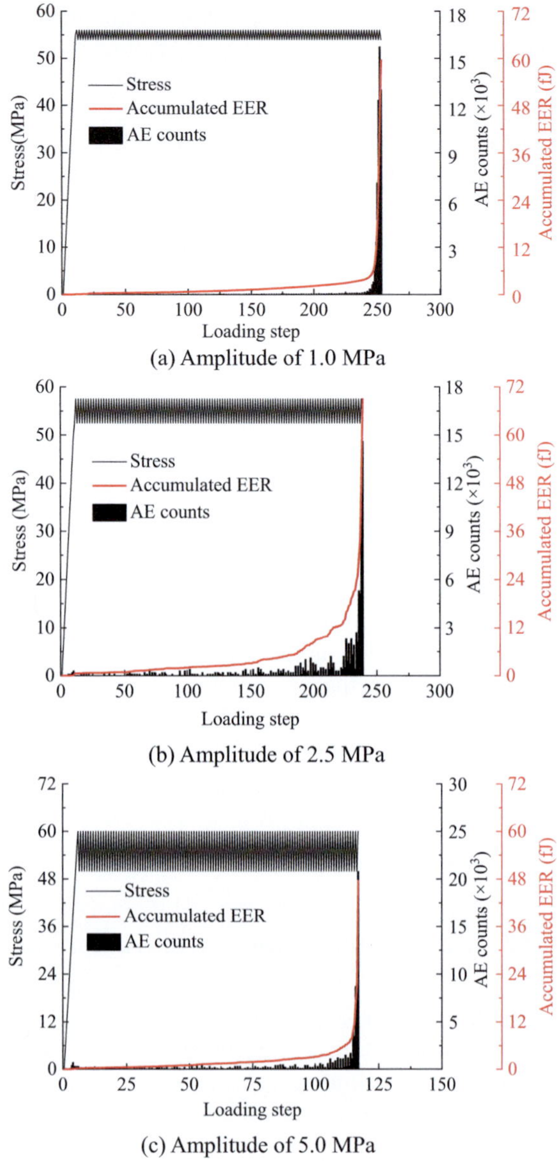

(a) Amplitude of 1.0 MPa

(b) Amplitude of 2.5 MPa

(c) Amplitude of 5.0 MPa

which P_h and P_v are respectively horizontal ground stress and vertical ground stress, is studied in this section by applying the newly developed numerical approach.

7.4.1.1 Model Configurations

Figure 7.10 shows a two-dimensional plane strain numerical model of an underground tunnel established in the new simulator. The quadrate model, which has dimensions of 50 m × 50 m, is excavated with a 10-m diameter hole. The mesh for the model has 500 × 500 = 250,000 quadrilateral elements of the same size. The elastic modulus and uniaxial compressive strength are considered to randomly vary among elements with the same homogeneity index ($m = 3$) by conforming to the Weibull distribution. The spatial assignment of elastic modulus to the mesoscale elements is shown in Fig. 7.10. The material properties and input parameters used in the model of the underground tunnels are presented in Table 7.4.

After the numerical model of the underground tunnel is established, it is pressurized at an original vertical ground stress P_v of 60.0 MPa in conjunction with

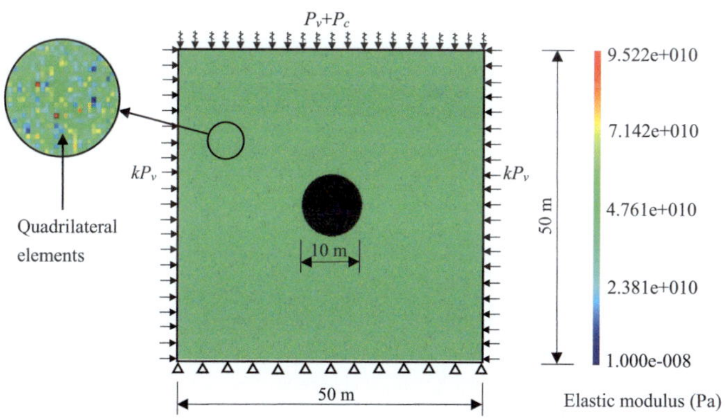

Fig. 7.10 A two-dimensional plane strain numerical model of an underground tunnel: spatial assignment of elastic modulus to the meso-scale elements

Table 7.4 Parameters employed in the numerical models of underground tunnels

Parameters	Values
Homogeneity index, *m*	3
Mean elastic modulus (GPa)	50
Poisson's ratio, *v*	0.25
Mean uniaxial compressive strength (MPa)	400
Friction angle (°)	30
Material parameter, *C*	23,259
Material parameter, *δ*	– 0.15549
Upper limit stress (MPa)	62.5
Lower limit stress (MPa)	57.5
Number of equivalent cycles per calculation step	100

horizontal ground stress of kP_v in the first calculation step. The lateral pressure coefficients k is specified as 0.2, 0.5, 1.0, 1.5, and 2.0, respectively, and five different numerical simulation analyses are conducted. Then, in subsequent loading steps, a cyclic loading P_c with a constant amplitude of 2.5 MPa is applied to the models in the vertical direction to investigate its influence on damage and failure response of the underground tunnels. After being superimposed with the vertical ground stress P_v, the models are subjected to a cyclic loading between 57.5 and 62.5 MPa.

7.4.1.2 Numerical Results

Figure 7.11a–c give the failure process of underground tunnels with five different lateral pressure coefficients. Figure 7.11a shows the maximum shear stress distribution around the underground tunnel under the original ground stresses before applying the cyclic load, corresponding to the first loading step. It is notable that, instead of quantitatively analyzing the magnitude of the maximum shear stress, the progressive damage is mainly qualitatively discussed in Fig. 7.11, so the color bars that mark values of the maximum shear stress are omitted. In Fig. 7.11, red represents the top values, blue represents the minimum values, and other colors represent the median values. From Fig. 7.11a, when $k = 0.2$ and $k = 0.5$, since the loads applied in the vertical direction P_v are more significant than the loads applied in the horizontal direction P_h, the areas with more significant maximum shear stress distribution appear on the left- and right-side walls of the underground tunnels. In contrast, the maximum shear stresses at the vault crown and inverted arch are more minor. When $k = 1.5$ and $k = 2.0$, the horizontal ground stress P_h is greater than the vertical ground stress P_v, resulting in more considerable maximum shear stress near the vault and invert and relatively more minor values at the left- and right-side walls. The positions where the maximum shear stresses are the largest in their model are located at the arch shoulders and feet of the tunnel that are approximately 45° from the horizontal and vertical directions of the model (it can also be seen from the AE distribution positions in the first loading step). For the case where the horizontal ground stress is equal to the vertical ground stress ($k = 1.0$), the maximum shear stress is evenly distributed in the circumferential direction of the tunnel. The area closest to the tunnel wall has the most considerable value of the maximum shear stress that gradually decreases to the initial stress of rock towards the far-field.

Figure 7.11b shows crack initiation of the underground tunnels with various lateral pressure coefficients k after the cyclic load in the vertical direction is applied. The ellipses circle the locations of crack initiation in the figures. When $k = 0.2$, the crack initiation position is at the top of the vault and the bottom of the invert, this is mainly due to the large difference between the vertical ground stress and the horizontal ground stress. Under the continuous action of cyclic loading–unloading, when the loading step 6 (approximately equivalent to 500 cycles of loading–unloading) is performed, the maximum shear stress at the left- and right-side walls has not reached its shear strength yet. Still, the tensile strength of the vault and the invert reaches its tensile strength under the indirect action of horizontal stretching, resulting in tensile

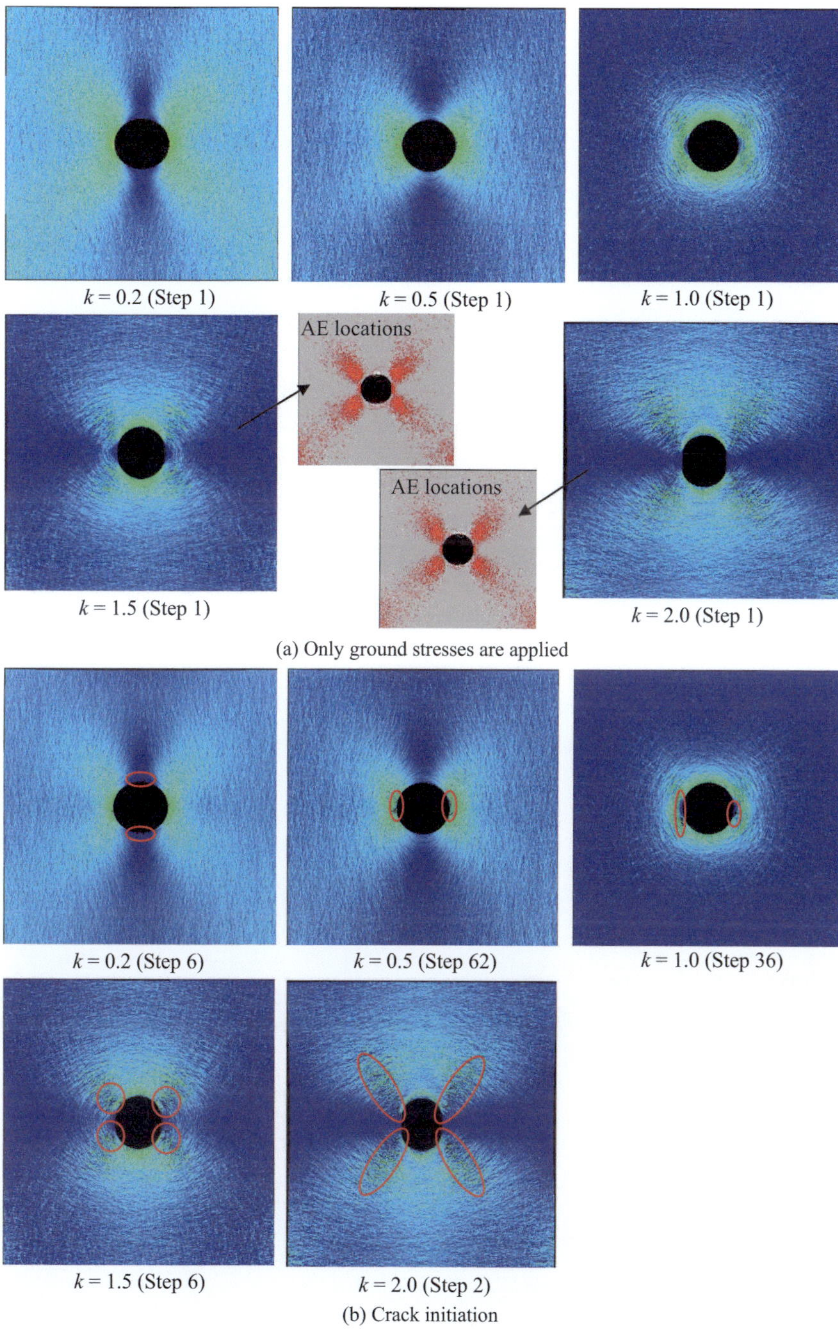

Fig. 7.11 Failure process of underground tunnels with various lateral pressure coefficients

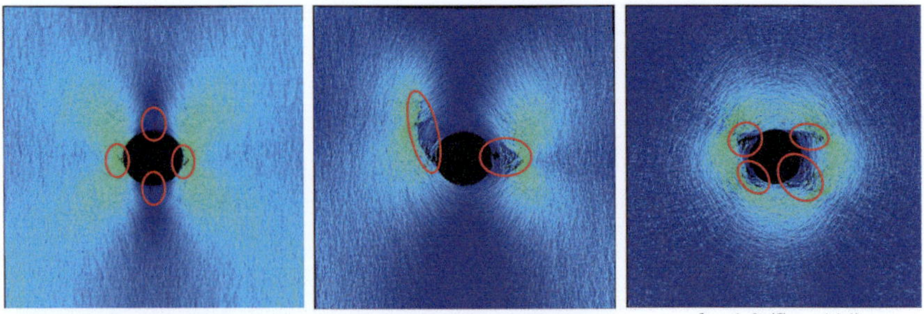

$k = 0.2$ (Step 62) $k = 0.5$ (Step 115) $k = 1.0$ (Step 114)

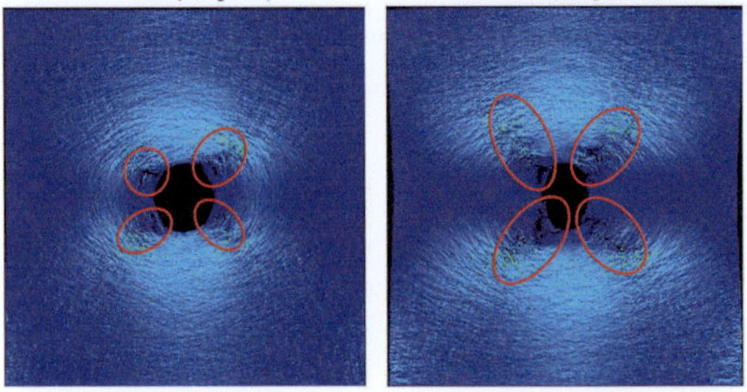

$k = 1.5$ (Step 56) $k = 2.0$ (Step 8)

(c) Crack propagation

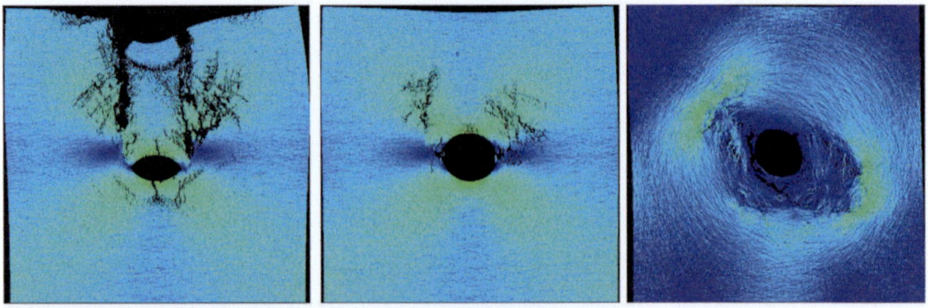

$k = 0.2$ (Step 89) $k = 0.5$ (Step 129) $k = 1.0$ (Step 164)

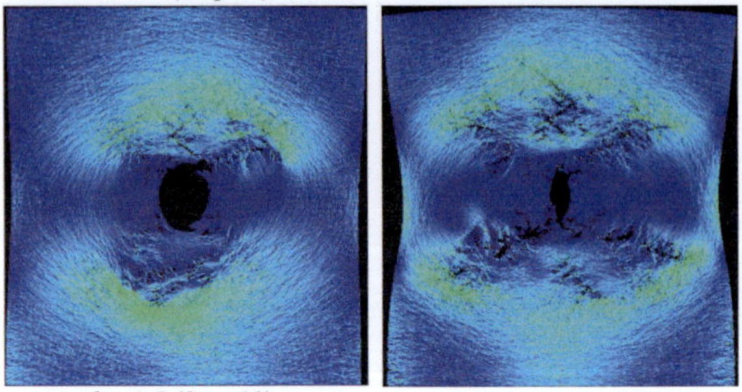

$k = 1.5$ (Step 89) $k = 2.0$ (Step 19)

(d) Failure

Fig. 7.11 (continued)

cracks. When k = 0.5, crack initiation occurs at loading step 62, corresponding to about 6200 cyclic disturbances, and the crack initiation position is at the left- and right-side walls of the tunnel. When the underground tunnel is under static ground stress, the maximum shear stress at the left- and right-side walls is relatively larger than in other areas. After the cyclic load is applied, the mesoscopic elements at these locations reach their damage thresholds, leading to shear cracks initiation. When $k = 1.0$, the crack initiation position is located at the tunnel's left- and right-side walls. When only the initial ground stresses are acted, the maximum shear stress of the tunnel is evenly distributed along the circumferential direction. After the cyclic load acts in the vertical direction, it breaks this distribution feature and finally causes the position of the left- and right-side walls to crack. The crack initiation, in this case, is at Step 36, roughly equivalent to 3500 cycles of loading–unloading. For the cases where the horizontal ground stress is greater than the vertical ground stress ($k = 1.5$ and $k = 2.0$), the maximum shear stress appears at the arch shoulders and feet of the tunnel under the initial ground stress. After the vertical cyclic load is applied, the crack initiation also appears in these same positions. The loading steps of crack initiation are Step 6 and Step 2, respectively, and the corresponding loading–unloading cycles are around 500 and 100 times. Hence, in terms of the number of cycles corresponding to crack initiation of the model, the number of loading–unloading cycles needed for $k = 0.5$ is the largest (about 6200 cycles), followed by $k = 1.0$ (about 3500 cycles), followed by $k = 0.2$ (about 500 cycles), and finally the case of $k = 1.5$ and $k = 2.0$ (approximately 500 and 100 cycles, respectively). From these data, we can conclude that the number of loading–unloading cycles corresponding to the initiation crack does not change with the lateral pressure coefficient k with an obvious regularity because the crack initiation of the tunnels in all these cases does not occur at the same positions around the arch. The types of cracks are also different, which can be divided into tensile cracks and shear cracks.

After a crack appears in the surrounding rock of the tunnel, the crack will continue to propagate under the continuous cyclic loading–unloading condition. Figure 7.11c shows the crack morphology characteristics circled by the red ellipses during crack propagation at a typical loading step. For $k = 0.2$, when the cyclic loading reaches loading step 62, the cracks at the top of the vault and the bottom of the invert start to expand slightly in the vertical direction, and cracks at the left- and right-side walls appear diagonally upward. For $k = 0.5$, the cracks on the left- and right-side walls also propagate in the same direction as $k = 0.2$, i.e., diagonally upward. From Fig. 7.11d, the failure mode of the tunnels for $k = 0.2$ and $k = 0.5$ attributes to the collapse of the vault downward; that is, the surrounding rock directly above the tunnels intrudes into the opening and fails. For $k = 1.0$, due to the horizontal and vertical ground stress being equal, despite the small amplitude (2.5 MPa) of the vertical cyclic loading, the vertical and horizontal stresses acting on the surrounding rock of the tunnel are still not much different. The crack propagation is mainly near the tunnel wall, and the shear stress still presents a relatively uniform distribution in the circumferential direction. The final failure is caused by the extrusion of the surrounding rock into the tunnel wall, forming a specific range of damage zone. For $k = 1.5$ and $k = 2.0$, the cracks at the arch shoulders and feet of the tunnel gradually expand along their crack initiation direction, which is about $\pm$ 45° in the vertical direction, forming the failure mode in the shape of an 'X'. At the same time, because the horizontal ground stress is greater than the vertical ground stress, the tunnels are mainly manifested as the left- and right-side walls squeezing into the tunnel when the cyclic loading triggers complete failure.

The above analysis can be roughly summarized as follows: when the vertical ground stress around tunnels is greater than the horizontal ground stress, the failure mode is mainly the collapse of the vault downward when tunnels are disturbed by a vertical cyclic loading with a relatively small amplitude; when the vertical ground stress is equal to the horizontal ground stress, the tunnel will form a damage zone with a certain thickness when subjected to the vertical cyclic loading; when the vertical ground stress is less than the horizontal ground stress, the failure of the tunnels is mainly caused by the extrusion of the left- and right-side walls into the opening. Therefore, although the tunnels achieve stability under the initial ground stress, the subsequent continuous cyclic load can also trigger progressive damage and even lead to failure and instability of the surrounding rock. Specifically, after the cyclic load is continuously applied to a certain extent in the vertical direction, all the tunnels with different lateral pressure coefficients will have crack initiation and propagation at different positions and in different directions, and the final failure modes of the tunnels are mainly determined by the ratio of initial horizontal to vertical ground stress.

7.4.2 Comparison with Previous Studies

Similar numerical approaches for rock materials subjected to cyclic loading were proposed by Song et al. (2019b) and Xu et al. (2021). Figure 7.12 illustrates numerical results of the stress–strain curves from previous studies. The numerical and experimental failure modes of the sandstone specimens from Xu et al. (2021) are also plotted in Fig. 7.12. It can be seen that the stress-loading step curves shown in Fig. 7.6 and the numerical failure modes shown in Fig. 7.7 in this research, besides of hysteresis loops in the stress–strain curves, were similar to previous studies.

However, the calculation amount of the large numbers of cyclic loading is much heavier in previous studies than that in this research. The parameter in the novel code—number of equivalent cycles per calculation step, can reduce the heave calculation amount to a great extent. This parameter means that the progressive damage of the mesoscopic elements in the numerical model induced by several dozens or even hundreds loading cycles can be can be regarded as an equivalent damage that occurs in a single loading step. For fatigue tests of rock sample, especially for high-cycle fatigue tests, it may take thousands of cycles of loading to cause fatigue failure. The previous method reported by Song et al. (2019b) and Xu et al. (2021) may need several loading steps to obtain a single hysteresis loop, but the method in this study has advantage of reducing the calculation amounts by several orders of magnitude. In addition, the damage progressivity of rock subjected to cyclic loading was rarely

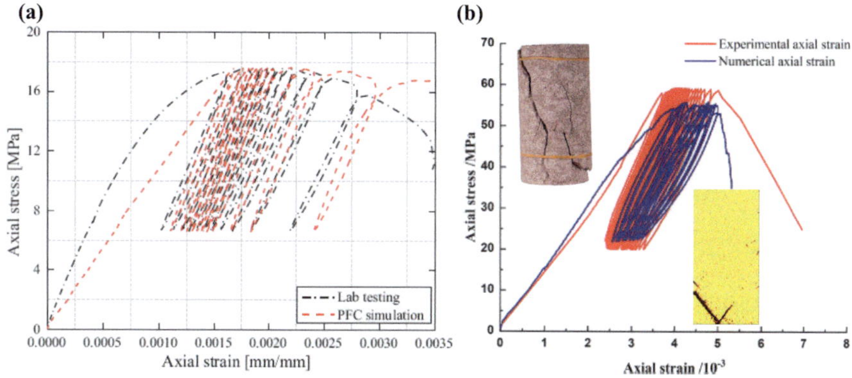

Fig. 7.12 Some numerical results with respect to stress–strain curves: **a** Song et al. (2019b) and **b** Xu et al. (2021)

reported in numerical simulation. The detailed evolutionary processes of damage for rock material or rock structures subjected to cyclic loading were deeply discussed, which is critical significance for understanding the crack propagation and fatigue failure mechanism (Khan et al. 2018). Furthermore, a technique to obtain information of damage location and damage energy corresponding to the AE monitoring in the laboratory or the microseismic monitoring in the field was seldom numerically simulated on numerical model of rock materials. The numerical approach in this study emphases the process of crack initiation, propagation and coalescence of rock under cyclic loading. The monitoring of crack location and energy conducted on the mesoscopic elements makes for revealing the crack propagation position, expanded direction, and energy characteristics of rock subjected to cyclic loading.

7.4.3 *Limitations*

7.4.3.1 Without Consideration of the Nonlinear Relationship of Strength Deterioration

As described in Sect. 7.2.2, the strength deterioration curve is simplified as a linear relationship (from point C to point C′ in Fig. 7.3) instead of some other nonlinear functions, even though some non-linear function forms between strength degradation or damage degree and cyclic number (Xiao et al. 2009; Liu and He 2012) have been proposed in the available literature. There are two main reasons for without considering the nonlinear relationship of strength deterioration: first, the non-linear models, which were empirically summarized based on experimental data rather than rigorous mathematical derivation, have not been widely accepted so far; second, the cross-scale correspondence between damage evolution of rock samples on the macroscopic scale and strength degradation of each element on the mesoscopic scale has not been established. Hence, notwithstanding the assumption of a linear relationship between strength degradation and cyclic number, the simplification practice is easy to use and is suitable for simulation for rock subjected to cyclic loading to some extent. More experimental and theoretical studies will be required in the future to establish a precise strength degradation model of mesoscopic elements for deformation, damage, and failure prediction in rock mechanics experiments and rock engineering.

7.4.3.2 Without Consideration of the Waveform Effect of Cyclic Loading

The waveform effect is not taken into consideration in the newly proposed model, which is another limitation of this work, even though some previous studies concluded that the behaviors of rocks, such as strength, damage evolution, fatigue life, damage accumulation, are susceptible to the waveform of the cyclic load (Tao and Mo 1990; Bagde and Petroš 2005). It does not mean that specific waveforms cannot be input into the code based on RFPA2D; on the contrary, some studies on waveform effect have been reported already. For example, Zhu et al. (2010) and Wang et al. (2012) conducted finite element analyses on rockburst of underground opening and failure of unreinforced masonry walls triggered by dynamic disturbance by applying a waveform of some trapeziform stress pulses on the models using a developed RPFA2D code. Zhu et al. (2012) also input a ramp waveform of stress pulse into RFPA2D to simulate the stress distribution and damage evolution in rock specimens during conventional split Hopkinson pressure bar (SHPB) tests.

The reason why the influence of the waveform is not considered in this study is to shorten the calculation steps. For cyclic loading–unloading tests or fatigue tests in rock mechanics, the number of cyclic loadings that the rock sample can withstand is often huge, especially at relatively lower stress levels $(\sigma_{max} + \sigma_{min})/2$ and lower amplitude of cyclic loading, which can reach tens of thousands or even hundreds of thousands of cycles. For instance, Burdine (1963) subjected Berea sandstone to cyclic uniaxial and triaxial compression and found that, at 74% of the monotonic (or static) compressive strength, failure occurred before 10^6 cycles. If the propagation (transmission and reflection) of waves in the rock sample in each loading cycle is considered, the calculation process will become highly complicated and time-consuming, and it will be challenging to obtain a good simulation result of damage evolution.

7.5 Chapter Summary

This chapter proposed a novel numerical model for simulating damage and failure of rock materials subjected to cyclic loading based on the theory of heterogeneity consideration and statistical damage constitutive model. The model's fundamental was the innovatively presented assumption that the strength of each mesoscopic element degrades with cyclic numbers according to a linear function. The experimental results in Chap. 4 were used to verify and validate the newly proposed model to simulate progressive damage and failure of rock subjected to cyclic loading. The numerically simulated results were tuned to the results of laboratory experiments.

Underground tunnels may be subject to cyclic disturbance load after excavating and stabilizing. To explore the instability process of underground tunnels induced by the remote vertical cyclic loading, numerical simulation of damage evolution and failure modes around underground tunnels for five different levels of lateral pressure coefficients were carried out using the developed numerical approach. To conclude, the continuous cyclic loading can trigger progressive damage and even lead to failure and instability of the tunnel surrounding rock. The final failure modes are highly susceptible to the lateral pressure coefficients.

The developed model has limitations. The strength deterioration curve is assumed to obey a simple linear function rather than other nonlinear functions, and the cyclic loading waveform effect is not considered. Despite these limitations, the model considering strength degradation of mesoscopic finite elements may still be a potentially powerful tool for understanding the progressive damage and failure mechanism of rock subjected to cyclic loading.

Suggestions for further work include validating the proposed two-dimension numerical model for rock damage and failure under other cyclic loadings, such as seismic and repeated traffic loads. Moreover, extending the current two-dimension model into a three-dimension will be carried out to explore the damage evolution and failure characteristics of underground tunnels or other rock structures subject to out-of-plane cyclic loadings.

References

Bagde MN, Petroš V (2005) Waveform effect on fatigue properties of intact sandstone in uniaxial cyclical loading. Rock Mech Rock Eng 38(3):169–196. https://doi.org/10.1007/s00603-005-0045-8

Bao CY, Jiang AN, Tang CA et al (2011) Study of acoustic emission characteristics of limestone under cycle uniaxial loading- unloading perturbation. Chin J Rock Mech Eng 30(S2):3871–3877 (in Chinese)

Broek D (1988) The practical use of fracture mechanics. Kluwer Academic Publishers, Dordrecht

Burdine NT (1963) Rock failure under dynamic loading conditions. Soc Petrol Eng J 3(1):1–8. https://doi.org/10.2118/481-PA

Dai B, Zhao GY, Konietzky H et al (2018) Experimental and numerical study on the damage evolution behaviour of granitic rock during loading and unloading. KSCE J Civ Eng 22(9):1–14. https://doi.org/10.1007/s12205-018-1653-7

Feng XT, Young RP, Reyes-Montes JM et al (2019) ISRM suggested method for in situ acoustic emission monitoring of the fracturing process in rock masses. Rock Mech Rock Eng 52(3):1395–1414. https://doi.org/10.1007/s00603-019-01774-z

Fu B, Hu LH, Tang CA (2020) Experimental and numerical investigations on crack development and mechanical behavior of marble under uniaxial cyclic loading compression. Int J Rock Mech Min Sci 130:104289. https://doi.org/10.1016/j.ijrmms.2020.104289

Khan AI, Venkataraman S, Miller I (2018) Predicting fatigue damage of composites using strength degradation and cumulative damage model. J Compos Sci 2(1):9. https://doi.org/10.3390/jcs2010009

Lemaitre J, Desmorat R (2005) Engineering damage mechanics: ductile, creep, fatigue and brittle failures. Springer, Berlin

Liang ZZ, Xing H, Wang SY et al (2012) A three-dimensional numerical investigation of the fracture of rock specimens containing a pre-existing surface flaw. Comput Geotech 45(9):19–33. https://doi.org/10.1016/j.compgeo.2012.04.011

Liu EL, He SM (2012) Effects of cyclic dynamic loading on the mechanical properties of intact rock samples under confining pressure conditions. Eng Geol 125(27):81–91. https://doi.org/10.1016/j.enggeo.2011.11.007

Liu MX, Liu EL, Feng JL et al (2021) Dynamic mechanical properties of jointed soft rock samples subjected to cyclic triaxial loading: a FEM-DEM-based study. IOP Conf Ser Earth Environ Sci 676(1):012066. https://doi.org/10.1088/1755-1315/676/1/012066

Ma SQ, He C, Zhao ZY et al (2017) Modeling of rock joints under cyclic loading conditions using discontinuous deformation analysis. Rock Mech Rock Eng 50(5):1205–1215. https://doi.org/10.1007/s00603-016-1158-y

Mlikota M, Schmauder S, Božić Ž (2018) Calculation of the Wöhler (S-N) curve using a two-scale model. Int J Fatigue 114:289–297. https://doi.org/10.1016/j.ijfatigue.2018.03.018

Song ZY, Konietzky H, Herbst M (2019a) Bonded-particle model-based simulation of artificial rock subjected to cyclic loading. Acta Geotech 14:955–971. https://doi.org/10.1007/s11440-018-0723-9

Song ZY, Konietzky H, Herbst M (2019b) Three-dimensional particle model based numerical simulation on multi-level compressive cyclic loading of concrete. Constr Build Mater 225:661–677. https://doi.org/10.1016/j.conbuildmat.2019.07.260

Tang CA (1997) Numerical simulation of progressive rock failure and associated seismicity. Int J Rock Mech Min Sci 34(2):249–261. https://doi.org/10.1016/S0148-9062(96)00039-3

Tang CA, Tham L, Lee P et al (2002) Coupled analysis of flow, stress and damage (FSD) in rock failure. Int J Rock Mech Min Sci 39(4):477–489. https://doi.org/10.1016/S1365-1609(02)00023-0

Tang CA, Wang JM, Zhang JJ (2010) Preliminary engineering application of microseismic monitoring technique to rockburst prediction in tunneling of Jinping II project. J Rock Mech Geotech Eng 2(3):193–208. https://doi.org/10.3724/SP.J.1235.2010.00193

Tang CA, Webb AAG, Moore WB et al (2020) Breaking Earth's shell into a global plate network. Nat Commun 11:3621. https://doi.org/10.1038/s41467-020-17480-2

Tao Z, Mo H (1990) An experimental study and analysis of the behaviour of rock under cyclic loading. Int J Rock Mech Min Sci Geomech Abstr 27(1):51–56. https://doi.org/10.1016/0148-9062(90)90008-P

Wang SY, Sloan SW, Abbo AJ et al (2012) Numerical simulation of the failure process of unreinforced masonry walls due to concentrated static and dynamic loading. Int J Solids Struct 49(2):377–394. https://doi.org/10.1016/j.ijsolstr.2011.10.016

Xi X, Yang S, Mcdermott CI et al (2021) Modelling rock fracture induced by hydraulic pulses. Rock Mech Rock Eng 54(8):3977–3994. https://doi.org/10.1007/s00603-021-02477-0

Xiao JQ, Ding DX, Xu G et al (2009) Inverted S-shaped model for nonlinear fatigue damage of rock. Int J Rock Mech Min Sci 46(3):643–648. https://doi.org/10.1016/j.ijrmms.2008.11.002

Xiao JQ, Ding DX, Jiang FL et al (2010) Fatigue damage variable and evolution of rock subjected to cyclic loading. Int J Rock Mech Min Sci 47:461–468. https://doi.org/10.1016/j.ijrmms.2009.11.003

Xiao JQ, Feng XT, Ding DX et al (2011) Investigation and modeling on fatigue damage evolution of rock as a function of logarithmic cycle. Int J Numer Anal Meth Geomech 35(10):1127–1140. https://doi.org/10.1002/nag.946

Xiao YX, Feng XT, Hudson JA et al (2016) ISRM Suggested method for in situ microseismic monitoring of the fracturing process in rock masses. Rock Mech Rock Eng 49(1):343–369. https://doi.org/10.1007/s00603-015-0859-y

Xu T, Fu M, Yang SQ et al (2021) A numerical meso-scale elasto-plastic damage model for modeling the deformation and fracturing of sandstone under cyclic loading. Rock Mech Rock Eng 54:4569–4591. https://doi.org/10.1007/s00603-021-02556-2

Yang TH, Tham LG, Tang CA et al (2004) Influence of heterogeneity of mechanical properties on hydraulic fracturing in permeable rocks. Rock Mech Rock Eng 37(4):251–275. https://doi.org/10.1007/s00603-003-0022-z

Zhu WC, Li ZH, Zhu L et al (2010) Numerical simulation on rockburst of underground opening triggered by dynamic disturbance. Tunn Undergr Space Technol 25(5):587–599. https://doi.org/10.1016/j.tust.2010.04.004

Zhu WC, Bai Y, Li XB et al (2012) Numerical simulation on rock failure under combined static and dynamic loading during SHPB tests. Int J Impact Eng 49:142–157. https://doi.org/10.1016/j.ijimpeng.2012.04.002

Chapter 8
Concluding Remarks and Prospects

8.1 Concluding Remarks

When excavation is carried out under high stress environment, the external disturbance will change the stress field of the surrounding rock, which may lead to potential engineering damage such as large deformation and large displacement of the free surfaces of the engineering surrounding rock, and even cause instability damage, rockburst and other engineering disasters. In order to find out the mechanical response and internal mechanism of damage and failure of rock triggered by dynamic disturbance, an electro-hydraulic servo triaxial test system for dynamic disturbance loading was developed and modified independently. Then, the modified equipment was used to carry out triaxial cyclic dynamic disturbance tests on red sandstone samples with different disturbance frequencies, amplitudes, and different levels of confining pressures. The characteristics of the stress–strain curves and stress-time curves of the rock samples, the influence of the frequency and amplitude of the cyclic dynamic disturbance on the mechanical response, the damage evolution during the disturbance process, the failure modes of the rock samples after loading, and the confining pressure effect were analyzed based on the experimental results. Subsequently, a statistical damage constitutive equation of rock in quasi-static loading phase was established, and the abrupt instability of the rock sample at the end of cyclic dynamic disturbance phase was explained using cusp catastrophe theory. Finally, a novel numerical model for simulating damage and failure of rock materials subjected to cyclic loading was proposed. And the experimental results were used to verify and validate the newly proposed model to simulate progressive damage and failure of rock subjected to cyclic loading. The following conclusions were drawn:

1. The frequency and amplitude of the axial disturbance load, and confining pressure have important effects on time duration of disturbance loading phase. Specifically, the time duration and the number of cycles were negatively correlated with the frequency and amplitude of cyclic dynamic disturbance, but were positively correlated with the confining pressure.

X. Chen, *Damage and Failure of Rock Triggered by Dynamic Disturbance*,
https://doi.org/10.1007/978-981-96-9486-0_8

2. The failure modes of rock samples under relatively lower frequencies and amplitudes coupling conditions characterize shear failure with a main inclined fracture and a few number of derivative cracks around; with the disturbance frequency and amplitude increase, the fragmentation degrees grow, the brittleness of failure weakens and the ductility strengthens; the failure modes under coupling conditions of relatively higher frequencies and amplitudes attribute cataclasis and exhibit some ductile characteristics. The sandstone samples were more broken and fragmented after the cyclic disturbance tests than after the conventional triaxial compression tests. As the confining pressure increased, the failure of the rock samples after the cyclic disturbance tests exhibited more severe flaking and exfoliation.
3. Damage variable defined by residual strain method, average dynamic axial stiffness method, and maximum axial and radial strain method can not only well reflect the gradual increase in damage degree of rock samples in disturbance loading tests under different disturbance frequencies and different amplitudes conditions, but can also characterize the initial damage degree of the disturbance loading phase.
4. According to the assumption that the strength of rock microbodies follows Weibull distribution and the failure conditions of the rock microbodies conform to the Hoek–Brown strength criterion, a damage evolution equation of the red sandstone sample in the quasi-static loading stage is established. The theoretical curve plotted by the constitutive equation have a good agreement with the experimental curves. The damage variable in the constitutive relationship shows an "S" shape with the evolution trend of axial strain, which accords with the characteristics of damage evolution of the rock samples under conventional triaxial compression.
5. Based on the cusp catastrophe theory and the mechanical model of the system of rock-test machine, a bifurcation set of the sudden jump of stress at the end of cyclic disturbance phase can be analytically obtained, which make a potential theoretical explanation for sudden instability or failure of rock subjected to cyclic disturbance.
6. With the assumption that the strength of each mesoscopic element degrades with cyclic numbers according to a linear function, a novel numerical model for simulating damage and failure of rock materials subjected to cyclic loading based on the theory of heterogeneity consideration and statistical damage constitutive model are proposed. The validation of the proposed numerical implementation method is verified by embedding it into a the RFPA2D code.

8.2 Prospects

Many factors are involved in rock damage and failure triggered by dynamic disturbances. The external factors include frequency, amplitude, waveform, location, and loading rate of the cyclic dynamic disturbance, and the internal factors include lithology, hardness, water content, stress state, and strain energy of the rock. If the rock is extended to the rock mass, factors such as characteristics of bedding, fracture (joint) distribution and structural plane are also included. In addition, the cross-consideration of these factors increases the difficulty of the research on damage and failure of rock triggered by dynamic disturbance. After preliminary research, the author believes that the problems that can be further discussed mainly include three aspects:

1. Experimental study

Firstly, to improve the obtained test data and clarify the influence of each factor on the test results, the influence of internal and external factors should be comprehensively considered in the tests. On the basis of the tests considering a single external or internal factor, the tests of multiple factors intersecting with each other should be carried out in the future.

Secondly, some more advanced test technologies should be introduced into the tests to visualize the damage and failure process of rock triggered by dynamic disturbance. Acoustic emission signal detection technology, computed tomography technology, infrared radiation capture technology, acoustic non-destructive testing technology, etc., have gradually begun to play a significant role in the experimental research on the evolution process of rock damage. By introducing these experimental techniques, the process can be observed from macroscopic and microscopic scales, and the damage and failure mechanism can be deeply explored.

In addition, the coupling problems of multiple fields (geostress field, temperature field, seepage field, chemical field, etc.) often involved in deep rock engineering should be taken into account in the next step of the tests.

2. Theoretical research

The numerical model of rock damage induced by cyclic disturbance established in this study also has room for improvement. Firstly, the disturbance frequency cannot be considered in this model. Secondly, the strength of each element in this model deteriorates continuously in a linear function under cyclic dynamic disturbance, and the equation is relatively simple and easy to use, but it cannot fully and accurately reflect the strength deterioration process. In future studies, this equation can be further modified to simulate the evolution process of rock damage triggered by cyclic dynamic disturbance more accurately.

The theory of rock damage and failure triggered by cyclic dynamic disturbance should be further improved. In future theoretical research, numerical methods such as discrete element method, boundary element method and finite difference method can be used to reproduce the rock damage and failure test process and engineering disasters under high stress conditions triggered by cyclic dynamic disturbance.

Also, more detailed test results can be used to summarize some other experimental laws that are more in line with engineering practice, and a rigorous and recognized rock constitutive equation can be established. On the basis of the established constitutive equation, the damage evolution equation and failure criteria of rock can be considered. The results of these mathematical and mechanical extrapolations can also be used to improve the corresponding numerical methods and compile relevant engineering software.

3. Engineering application

The problem of damage and failure of rock triggered by dynamic disturbance can be divided into two aspects: harmful and beneficial for practical engineering applications. On the harmful side, the influence of external disturbance, from the perspective of safety and stability, should be fully considered in the design and construction of rock engineering. Relevant monitoring, early warning and preventive measures should be taken in advance for the surrounding rock that may be disturbed to trigger instability. Surrounding rock in need of support should be strengthened with sufficient strength in time, and measures should be taken in advance for the location where stress release is required. Also, the frequency, amplitude, and energy of the cyclic dynamic disturbance should be strictly controlled to avoid engineering accidents and ensure the safety of personnel and property. On the beneficial side, how to make the process of rock failure triggered by external dynamic disturbance be used should be considered. For example, a disturbance module can be added to develop the rock breaking machinery and to achieve the purpose of accelerating rock breaking. By means of controlling the frequency, amplitude and waveform of the cyclic dynamic disturbance precisely, the rock breaking efficiency can be improved on the premise of ensuring construction safety.

MIX
Papier aus verantwortungsvollen Quellen
Paper from responsible sources
FSC® C105338

If you have any concerns about our products,
you can contact us on
ProductSafety@springernature.com

In case Publisher is established outside the EU,
the EU authorized representative is:
Springer Nature Customer Service Center GmbH
Europaplatz 3, 69115 Heidelberg, Germany

Printed by Libri Plureos GmbH
in Hamburg, Germany